T0326260

correlations but, sometimes, simple reasoning is enough to account for a phenomenon.

Since our very beginnings on this planet, humans have had to deal with the four primordial "elements" as they were known in the ancient world: earth, water, air and fire (and a fifth: aether). Today, we speak of gases, liquids, minerals and vegetables, and finally energy.

The unit operation expressing the behavior of matter are described in thirteen volumes.

It would be pointless, as popular wisdom has it, to try to "reinvent the wheel" – i.e. go through prior results. Indeed, we well know that all human reflection is based on memory, and it has been said for centuries that every generation is standing on the shoulders of the previous one.

Therefore, exploiting numerous references taken from all over the world, this series of books describes the operation, the advantages, the drawbacks and, especially, the choices needing to be made for the various pieces of equipment used in tens of elementary operations in industry. It presents simple calculations but also sophisticated logics which will help businesses avoid lengthy and costly testing and trial-and-error.

Herein, readers will find the methods needed for the understanding the machinery, even if, sometimes, we must not shy away from complicated calculations. Fortunately, engineers are trained in computer science, and highly-accurate machines are available on the market, which enables the operator or designer to, themselves, build the programs they need. Indeed, we have to be careful in using commercial programs with obscure internal logic which are not necessarily well suited to the problem at hand.

The copies of all the publications used in this book were provided by the *Institut National d'Information Scientifique et Technique* at Vandœuvre-lès-Nancy.

The books published in France can be consulted at the *Bibliothèque Nationale de France*; those from elsewhere are available at the British Library in London.

In the in-chapter bibliographies, the name of the author is specified so as to give each researcher his/her due. By consulting these works, readers may

Preface

The observation is often made that, in creating a chemical installation, the time spent on the recipient where the reaction takes place (the reactor) accounts for no more than 5% of the total time spent on the project. This series of books deals with the remaining 95% (with the exception of oil-fired furnaces).

It is conceivable that humans will never understand all the truths of the world. What is certain, though, is that we can and indeed must understand what we and other humans have done and created, and, in particular, the tools we have designed.

Even two thousand years ago, the saying existed: "faber fit fabricando", which, loosely translated, means: "*c'est en forgeant que l'on devient forgeron*" (a popular French adage: *one becomes a smith by smithing*), or, still more freely translated into English, "practice makes perfect". The "artisan" (faber) of the 21st Century is really the engineer who devises or describes models of thought. It is precisely that which this series of books investigates, the author having long combined industrial practice and reflection about world research.

Scientific and technical research in the 20th Century was characterized by a veritable explosion of results. Undeniably, some of the techniques discussed herein date back a very long way (for instance, the mixture of water and ethanol has been being distilled for over a millennium). Today, though, computers are needed to simulate the operation of the atmospheric distillation column of an oil refinery. The laws used may be simple statistical

Contents

First published 2016 in Great Britain and the United States by ISTE Press Ltd and Elsevier Ltd

ISTE Press Ltd
27-37 St George's Road
London SW19 4EU
UK

www.iste.co.uk

Elsevier Ltd
The Boulevard, Langford Lane
Kidlington, Oxford, OX5 1GB
UK

www.elsevier.com

Notices

Knowledge and best practice in this field are constantly changing. As new research and experience broaden our understanding, changes in research methods, professional practices, or medical treatment may become necessary.

Practitioners and researchers must always rely on their own experience and knowledge in evaluating and using any information, methods, compounds, or experiments described herein. In using such information or methods they should be mindful of their own safety and the safety of others, including parties for whom they have a professional responsibility.

To the fullest extent of the law, neither the Publisher nor the authors, contributors, or editors, assume any liability for any injury and/or damage to persons or property as a matter of products liability, negligence or otherwise, or from any use or operation of any methods, products, instructions, or ideas contained in the material herein.

For information on all our publications visit our website at http://store.elsevier.com/

British Library Cataloguing-in-Publication Data
A CIP record for this book is available from the British Library
Library of Congress Cataloging in Publication Data
A catalog record for this book is available from the Library of Congress
ISBN 978-1-78548-179-6

Printed and bound in the UK and US

Industrial Equipment for Chemical Engineering Set

coordinated by
Jean-Paul Duroudier

Adsorption – Dryers for Divided Solids

Jean-Paul Duroudier

ELSEVIER

There are no such things as applied sciences,
only applications of science.
Louis Pasteur (11 September 1871)

Dedicated to my wife, Anne, without whose unwavering support, none of this
would have been possible.

Adsorption – Dryers for Divided Solids

gain more in-depth knowledge about each subject if he/she so desires. In a reflection of today's multilingual world, the references to which this series points are in German, French and English.

The problems of optimization of costs have not been touched upon. However, when armed with a good knowledge of the devices' operating parameters, there is no problem with using the method of steepest descent so as to minimize the sum of the investment and operating expenditure.

Humid Air Water Cooling Towers

1.1. Properties of humid air

1.1.1. *Absolute humidity and relative humidity*

We shall base our discussion here on the most common case, which is that of a mixture of air and water vapor. The conclusions drawn can easily be transposed to apply to gaseous mixtures of a different nature.

The absolute humidity of an air–water vapor mixture is written as Y. It is the mass (in kg) of water vapor associated with 1 kg of dry air.

The air is saturated with vapor when the partial pressure of the vapor P_V is equal to the vapor pressure π (t) corresponding to the temperature of the mixture.

The relative humidity is the ratio:

$$\varepsilon = \frac{P_V}{\pi(t)}$$

The saturation rate is the ratio of the absolute humidity Y of a given mixture to the humidity Y_s which it would have had if it were saturated.

The absolute humidity of humid air is easily calculated as a function of the partial vapor pressure P_V:

$$Y = \frac{18 \times P_V}{29 \times (P - P_V)}$$

We shall let "B" represent the ratio 18/28.94 = 0.622, and Y *is written:*

$$Y = \frac{B\,P_V}{P - P_V} = \frac{B\,\varepsilon\,\pi}{\left(P - \varepsilon\,\pi\right)} \tag{1.1}$$

1.1.2. *Enthalpy of humid air*

The enthalpy of water vapor at a given temperature can be determined in two different ways, which yield the same result.

By definition, the enthalpy of a pure substance is:

$$H = U + PV$$

U: internal energy.

In drying, we do not take account of the mechanical work due to the variations in the term PV, so any variation in enthalpy is due to an exchange of heat and the enthalpy of a pure vapor, along with the internal energy, depends solely on the temperature.

It is usual to take the origin of enthalpies at 0°C. Put differently, the enthalpy of water at 0°C is null, but this is simply a convention.

The enthalpy of water vapor at t°C is equal to the heat applied in one of the following two processes:

Vaporization of the water at 0°C and heating of the vapor to t°C;

Heating of the water to t°C and vaporization at t°C.

In the first case, the heat applied is: $L_0 + C_V t$.

In the second case, $C_E\,t + L_t$.

L_0 and L_t: latent heats of state-change of water at 0°C and at t°C.

These two values express the enthalpy of the water vapor in accordance with the principle of initial state and final state. For a function of state,

$$H_V = L_0 + \int_0^t C_V \, dt = \int_0^t C_E \, dt + L_t$$

The indices V and W refer respectively to the vapor and water.

In practical terms, the enthalpies of water and vapor are expressed, according to Cadiergues [CAD 78], by his equations 8 and 9:

$$H_E = \frac{t}{0.2374 + 4.015 \times 10^{-5} t - 2.721 \times 10^{-7} \, t^2}$$

$$H_V = 2500.8 + 1.8266 \, t + 2.818.10^{-4} t^2 - 1.0862.10^{-5} \, t^3$$

H_E and H_V : gravimetric enthalpies of water and vapor: J.kg^{-1};

t: temperature °C.

The gravimetric enthalpy of dry air varies little with the temperature:

$$H_G = 1006 \, t + 0.05 \, t^2 \qquad \left(J.kg^{-1} \right)$$

The enthalpy of humid air in relation to the mass of dry air is, finally:

$$H_{GH} = H_G + Y \, H_V \qquad\qquad [1.2]$$

When we know the temperature and enthalpy H_{GH}, it is easy to obtain Y.

[CAD 78] gives a general expression for H_{GH} (t, Y).

1.1.3. *Density of humid air*

Let P_{G0} be the density of dry air in normal conditions. The density of dry air still at the pressure of 1 bar, but at the absolute temperature T is, according to the Boyle–Mariotte law:

$$\rho_G = \rho_{G0} \frac{T_0}{T} = \frac{M_G P}{RT}$$

Now consider 1 m^3 of humid air. The molar fraction of the water vapor is:

$$y_V = \frac{Y/18}{1/29 + Y/18} = \frac{Y}{B+Y}$$

T: absolute temperature: K;

P: pressure: Pa;

Y: in humid air, the ratio of the mass of water vapor to the mass of dry air.

The mass of dry air occupying the volume V at the pressure P is:

$$m_G = M_G n = M_G \frac{PV}{RT}$$

The density of that air is, as we have just seen:

$$\rho_G = \frac{m_G}{V} = \frac{M_G P}{RT} \qquad [1.3]$$

Now consider 1 m^3 of humid air. The partial pressure of the dry air in that mixture at the total pressure P is P $(1 - y_V)$, and the mass of dry air present is:

$$m_G = \frac{M_G P (1 - y_V)}{RT} = \rho_G (1 - y_V)$$

The mass of humid air making up this mixture occupying 1 m^3 is:

$$\rho_{GH} = m_G (1 + Y) = \rho_G \left(1 - \frac{Y}{B+Y}\right)(1+Y) = \rho_G \frac{B(1+Y)}{B+Y} \qquad [1.4]$$

However:

$$\frac{B M_G}{R} = \frac{0.622 \times 29}{8314} = \frac{1}{461}$$

Thus, we can write:

$$\rho_{GH} = \frac{M_G P(1+Y)B}{RT(B+Y)} = \frac{P(1+Y)}{T \times 461(B+Y)} \qquad [1.5a]$$

As the molecular mass of a gas is proportional to its density, the mean molecular mass of the humid air is:

$$M_{GH} = M_G \frac{B(1+Y)}{(B+Y)} \qquad [1.5b]$$

The volume V_{GH} of humid air corresponding to 1 kg of dry air is:

$$V_{GH} = \frac{1+Y}{\rho_{GH}} = \frac{461}{P}(0.622+Y)T$$

1.1.4. *Vapor pressure of water*

The vapor pressure of water is given by:

$$\pi(t) = \exp\left[6.4145 + \frac{17.0057\,t}{233.08+t}\right]$$

π (t): pressure of the water vapor: Pa;

t: temperature of the water: °C.

1.1.5. *Transfer coefficients and Lewis number*

Remember that the Prandtl number Pr is $\dfrac{C\mu}{\lambda}$

C: specific heat capacity of the fluid: $J.\,kg^{-1}.^\circ C^{-1}$;

μ: dynamic viscosity of the fluid: Pa.s. $= kg.m^{-1}.s^{-1}$;

J: unit of work: $kg.m^2.s^{-2}$;

λ: heat conductivity of the fluid: $J.s^{-1}.^\circ C^{-1}.m^{-1}$.

The Schmidt number Sc is $\dfrac{\mu}{\rho D}$

ρ: density of the fluid: $kg.m^{-3}$;

D: diffusivity of the solute in the fluid: $m^2.s^{-1}$.

The Prandtl number and the Schmidt number are dimensionless, as is the Lewis number, which is the ratio $\dfrac{Pr}{Sc} = \dfrac{C\rho D}{\lambda}$.

Let us now consider operation *in the turbulent regime*. The heat transfer coefficient is of the form:

$$\alpha = \frac{\lambda}{\delta} \, j(Re) Pr^{1/3}$$

The material transfer coefficient is:

$$\beta = \frac{D}{\delta} \, j(Re) Sc^{1/3}$$

α: heat transfer coefficient: $W.m^{-2}.^\circ C^{-1}$;

β: material transfer coefficient: $m.s^{-1}$;

δ: reference length: m.

The reference length δ is the same for both transfers.

Experience shows us that, in the turbulent regime, the function j (Re) is the same for α and β. The result of this is that:

$$\frac{\alpha}{\beta} = \frac{\lambda}{D} \left(\frac{Pr}{Sc} \right)^{1/3} = \frac{\lambda}{D} Le^{1/3} = \frac{\lambda}{D} \left(\frac{C\rho D}{\lambda} \right)^{1/3}$$

If, as is the case for the water–air system, the Lewis number is equal to 1,

$$\frac{\lambda}{D} = C\rho \text{ and } \frac{\alpha}{\beta} = C\rho = C_{GH}\,\rho_{GH} \qquad [1.6]$$

C_{GH}: specific heat capacity of the humid gas: $J.kg^{-1}.°C^{-1}$.

1.1.6. *Wet-bulb thermometer equation*

Let us write that the cooling of the gas serves to vaporize the water in the sleeve surrounding the wet-bulb thermometer.

$$\alpha\left(t_G - t_H\right) = \beta \Lambda \left(C_H - C_{GH}\right)$$

α: heat transfer coefficient: $W.m^{-2}.°C^{-1}$;

Λ : molar latent heat of vaporization: $J.kmol^{-1}$;

β: material transfer coefficient: $m.s^{-1}$;

C_{GH}: concentration of the vapor in the gas: $kmol.m^{-3}$;

C_H: concentration of the vapor at equilibrium with the sleeve of the wet-bulb thermometer: $kmol.m^{-3}$.

Let us set:

L: latent heat of vaporization per kg of vapor: $J.kg^{-1}$;

C'_H and C'_G: concentration in mass: $kg.m^{-3}$.

According to equation [1.2]

$$C' = \frac{Y}{1+Y}\,\rho_{GH} = \rho_G\,\frac{YB}{B+Y}$$

The material transfer is written:

$$\beta\Lambda\left(C_H - C_{GH}\right) = \beta L\left(C'_H - C'_G\right) = \beta L \rho_G B\left(\frac{Y_H}{B+Y_H} - \frac{Y_{GH}}{B+Y_{GH}}\right)$$

The equation of the wet-bulb thermometer can be written thus, according to equation [1.6].

$$0 = \alpha\left(t_{GH} - t_H\right) + \beta\Lambda\left(C_{GH} - C_H\right)$$

$$= \beta\overline{C}_{GH}\overline{\rho}_{GH}\left(t_{GH} - t_H\right) + \beta L\rho_G B\left(\frac{Y_{GH}}{B + Y_{GH}} - \frac{Y_H}{B + Y_H}\right)$$

However, according to equation [1.5b]:

$$\rho_{GH} = \rho_G \frac{\left(1 + Y_{GH}\right)B}{\left(B + Y_{GH}\right)}$$

Finally:

$$0 = \beta\rho_G B\left[\frac{1 + Y_{GH}}{B + Y_{GH}}C_{GH}\left(t_{GH} - t_H\right) + L\left(\frac{Y_{GH}}{B + Y_{GH}} - \frac{Y_H}{B + Y_H}\right)\right] \qquad [1.7]$$

C_{GH}: specific heat capacity of the humid gas: $J.kg^{-1}.°C^{-1}$

or indeed, by setting $\left(1 + Y_{GH}\right)C_{GH} \# \overline{C}_{GH}$ and $B + Y_{GH} \# B + Y_H$

$$\overline{C}_{GH}\left(t_{GH} - t_H\right) + L\left(Y_{GH} - Y_H\right) = H_{GH} - H_H = 0 \qquad [1.8]$$

This equation is known as the wet-bulb thermometer equation.

The cooling of the air is compensated by its enrichment in water. The enthalpy of the air in contact with the wet-bulb thermometer does not vary.

1.2. Mollier diagram

In Europe, there is a tradition of representing the properties of humid air using the Mollier diagram [MOL 23].

On this diagram, established at atmospheric pressure, the ordinate axis shows the *enthalpy of the humid air* and the abscissa axis the *absolute humidity* Y of the air.

In addition, we plot the lines corresponding to constant values of the properties of the air:

1) Lines $\varepsilon = $ const.

We know that $\varepsilon = \dfrac{P_V}{\pi(t)}$

Conversely: $t_H = t(\pi) = f(P_V / \varepsilon)$

However,

$$P_V = \frac{P\,Y}{B + Y}$$

Thus:

$$t_H = f\left(\frac{P\,Y}{\varepsilon(B + Y)}\right)$$

2) Lines $P_V = $ const.

$$P_V = \frac{P\,Y}{B + Y}$$

These lines, therefore, are vertical.

3) Lines $t = $ const.

The enthalpy of the humid air expressed in relation to the dry air is written, in accordance with equation [1.2]:

$$H_{GH} = H_G + Y H_V$$

These lines are therefore straight lines whose ordinate at the origin is H_G and whose slope is H_V. As the enthalpy of the vapor includes its latent heat, that slope is steep.

4) Lines ρ_{GH} = const.

From the expression of ρ_{GH} (equation [1.4]), we derive that of t_{GH}, which is the temperature of the humid air:

$$t_{GH} = \frac{P}{\rho_{GH} \times 0.622} \times \frac{(1+Y)}{(0.622+Y)} - 273$$

The temperature decreases slightly when Y rises from zero. Apart from this restriction, the curves ρ_{GH} = const. are similar to the lines t = const.

5) Lines t_H = const. (constant humid temperature)

As the humid temperature is constant, the humidity Y_H is too, because these two properties are linked by the saturation (see equation [1.1], where $\varepsilon = 1$).

The wet-bulb thermometer equation (see equations [1.7] and [1.8]) can be used to calculate t_{GH} if we know t_H and Y_{GH}:

$$t_{GH} = t_H + \sigma \frac{L_{tH}}{C_{GH}} \left(Y_H - Y_{GH} \right) \qquad [1.9]$$

C_{GH} : specific heat capacity of the humid air: $J.kg^{-1o}C^{-1}$

t_{GH} : so-called "dry" temperature of the gaseous mixture: ^oC

$$C_{GH} = \frac{dH_{GH}}{dt}$$

We can easily deduce the enthalpy (see section 1.2).

The lines t_H = const. are curves with a slight positive slope. Indeed, we can write approximately:

$$H_{GH}(t_H) \# L_0 Y_{GH} + C_{GH} t_H + \sigma L \left(Y_H - Y_{GH} \right)$$

L is the mean value of the latent heat between t_{GH} and t_H

$$H_{GH}\left(t_H\right)\#C_{GH}t_H + \sigma\,LY_H + Y_{GH}\left(L_0 - \sigma\,L\right) = \text{const.} + Y_{GH}\left(L - \sigma\,L\right)$$

We can clearly see that the coefficient of Y_{GH} is low, because σ is nearly equal to 1 for the air–water system.

To increase the accuracy, it is customary to establish the Mollier diagrams in such a way that the two axes form an angle ranging, depending on the case, between 120° and 170°, so that the lines t_H = const. become practically horizontal.

A diagram for humid air is to be found at the end of Dascalescu's book [DAS 69]. [MOL 23] gives a smaller-scale model.

As the enthalpies H_G and H_{GH} and the humidity Y are relative to the unit mass of dry air, the mixing law applies, provided that each component is weighted by its mass of dry air. In the diagram, *the mixture is the barycenter* of the images of the components. This property is useful in performing the calculations for a dryer where the air is recycled.

1.3. Measurement of the humid temperature

1.3.1. *Wet-bulb thermometer and psychrometer*

The wet-bulb thermometer serves to measure the humid temperature t_H of the air. Its bulb is surrounded by a wick that is immersed in the water. Therefore, that wick is constantly soaked with water.

Figure 1.2 illustrates this device.

The psychrometric gap is the difference between the dry temperature t_{GH} and the humid temperature t_H :

$$t_{GH} - t_H = \frac{\sigma\,L}{C_P}(Y_H - Y_{GH})$$

Remember that:

$$Y_H = \frac{B\, y_H}{1 - y_H}; \quad y_H = \frac{\pi_H}{P}; \quad \sigma = (Lu)^{2/3}$$

For the air–water system, the coefficient σ is very close to 1, on condition that the Chilton–Colburn analogy is verified, meaning that we are in the turbulent regime.

Lu: Luikov number. This number is very close to 1 for water vapor in air. The Luikov number is none other than the Lewis number.

π_H : vapor pressure of water at t_H: Pa.

L: latent heat of vaporization of water from the liquid state (at temperature t_H) to the vapor state (at temperature t_G): J.kg^{-1}.

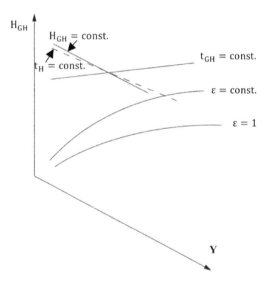

Figure 1.1. *Shape of the Mollier diagram*

A psychrometer is a device made up of two temperature probes:

– one measures the dry temperature t_{GH};

– the other measures the humid temperature t_H.

In order for the measurement of t_H to be correct, the Chilton–Coburn analogy must be valid, meaning that we must be in the turbulent regime around the probe. It is recommended that the gaseous flow rate be greater than 2 m.s^{-1}.

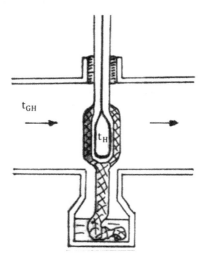

Figure 1.2. *Wet-bulb thermometer*

1.4. The cooling tower

1.4.1. *Description of the tower*

As regards the internal packing of the tower, in the past, stacks of horizontal planks of wood were used. Each layer was laid at a right angle from the layers above and below it. However, this device was rather heavy, and the surface usefully wetted by the water did not surpassed 30% of the total surface area of the planks.

Today, plates of PVC (polyvinyl chloride) are used, arranged vertically and parallel to one another on a given level. On a level, the vertical plates are at a right angle to the plates on the levels above and below it.

Thus, at least 70% of each face of a plate is wetted by the liquid. In addition, these plates are significantly lighter than the wooden planks, which produces a saving in terms of the construction of the tower.

The tower is often in the form of a hyperbolic paraboloid. Any vertical section passing through the axis is a hyperbola. That surface is engendered by a straight line inclined in relation to an axis, situated at a certain distance from that axis and rotating around it. This configuration ensures the best possible mechanical strength to support the internal packing, as well as the weight of the water which runs off.

As regards the partial coefficient α_L of heat transfer on the side of the liquid, readers can refer to the discussion in section 3.3.5 of [DUR 16b]. We would do the same for the coefficient β_G of material transfer on the side of the air.

1.4.2. *Calculation of the tower*

The equilibrium curve, also known as the saturation curve, is calculated as follows.

The partial pressure of the water vapor is:

$$P_V = P y_V = \frac{P\,Y}{B+Y} \text{ or indeed } Y = \frac{P_V B}{P - P_V}$$

At the air–water interface, the equilibrium is established so that:

$$P_V = \pi(t) \text{ and } Y = \frac{B\pi(t)}{P - \pi(t)} = Y(t)$$

Referring to development 1.1.2, the equation of the equilibrium curve (E) is therefore:

$$H(t) = H_G(t) + Y(t) H_V(t)$$

The slope of the operating line is obtained by writing that the heat lost by the water is gained by the air.

$$W_E\, C_E\, d\, t_E = W_G\, dH_{GH}$$

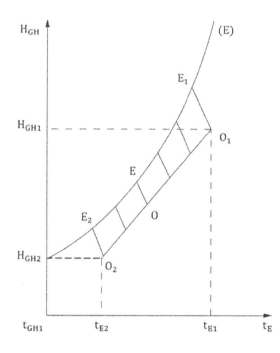

t_{E1}:   temperature of incoming water
t_{E2}:   temperature of outgoing water
t_{GH1}:   wet-bulb temperature of incoming air

the lines such as OE are the correspondence lines

the line O_1O_2 is the operating line

the curve (E) is the equilibrium   curve $H = f(t)$

the index 2 refers to the bottom of the tower and the index 1 to the top of the tower.

Figure 1.3. *Graphical construction for the numerical calculation of the height of the tower*

Hence:

$$\frac{dH_{GH}}{d\,t_E} = \frac{W_E C_E}{W_G} = \text{const.}$$

The temperatures of the water at inlet and outlet t_{E1} and t_{E2} are known, as is the humid temperature of the exiting air t_{GH1}, and consequently, so too is the enthalpy H_{GH2} of the incoming air.

The enthalpy H_{GH1} of the exiting air is given by the overall heat balance of the tower.

$$\frac{H_{GH1} - H_{GH2}}{t_{E1} - t_{E2}} = \frac{W_E\ C_E}{W_G}$$

Thus, the operating line is perfectly determined.

By writing that the density of heat flux lost by the water is gained by the air, we obtain:

$$\beta'_G \left(H_I - H_{GH} \right) = - \alpha_L \left(t_I - t_E \right)$$

The index I represents the water–air interface situated on the equilibrium curve (E). The values H_{GH} and t_E characterize the air and water at a given level in the column. On the basis of the point (H_{GH}, t_E) on the operating line, it is possible to determine the point representing the interface on the curve (E) by writing the slope of the correspondence line OE.

$$\frac{H_I - H_{GH}}{t_I - t_E} = - \frac{\propto_L}{\beta'_G} \qquad [1.10]$$

We know α_L (see section 3.3.5 of [DUR 16b]), but we need to calculate β'_G.

According to equation [1.9], the equation for the dimensions of the coefficient β'_G is:

$$\left[\beta'_G \right] = \frac{kg}{sm^2} = \frac{m}{s} \times \frac{kg}{m^3} = \left[\beta_G \right] \times \left[c \right]$$

In the gaseous phase, the driving force must be expressed as being a difference between the concentrations for water vapor.

The concentration of the water vapor in the humid gas is:

$$c_{VGH} = \frac{P_V M_V}{R\,T} = \frac{P\,M_V}{R\,T} \frac{Y_{GH}}{B + Y_{GH}} \quad as\ P_V = \frac{P\,Y}{B + Y}$$

M_V : molar mass of the water vapor: $kg.kmol^{-1}$

The concentration at the interface is:

$$c_I = \frac{P\,M_V\,Y_I}{R\,T(B + Y_I)} \quad \left(kg.m^{-3} \right)$$

β_G is then measured in m.s^{-1}, which is the traditional unit for measuring that transfer coefficient. Therefore, it is calculated by using the usual correlations (see section 3.3.5 of [DUR 16b]).

Finally, the flux density of vapor from the liquid to the gas is:

$$N_V = \beta_G \frac{P M_V}{R T} \left(\frac{Y_I}{B + Y_I} - \frac{Y_{GH}}{B + Y_{GH}} \right) = \beta \frac{P M_V}{R T} \left(Y_I - Y_{GH} \right) \left(kg.m^{-2}.s^{-1} \right)$$

The latent heat provided to the gas is:

$$L N_V = \frac{\beta_G \; L \; P \; M_V}{R T} \left(\frac{Y_I}{B + Y_I} - \frac{Y_{GH}}{B + Y_{GH}} \right)$$

L: latent heat of vaporization of the water: J.kg^{-1}.

The heat transfer coefficient α_G is given by:

$$\alpha_G = C_{GH} \beta_G' = C_{GH} \; x \; c \; x \; \beta_G = C_{GH} \; x \; \beta_G \; x \frac{P M_V}{R T}$$

C_{GH} : specific heat capacity of the humid gas: J.kg^{-1o}C^{-1}.

Finally, the heat flux density from the water to the gas is:

$$\Phi = \frac{\beta_G \; P \; M_V}{R T} \left[C_{GH} \left(t_I - t_{GH} \right) + L \left(\frac{Y_I}{(B + Y_I)} - \frac{Y_{GH}}{B + Y_{GH}} \right) \right]$$

Note the similarity with equation [1.7].

The term in square brackets is very close to the difference in enthalpies:

$$\left(H_I - H_{GH} \right)$$

Thus:

$$\Phi = \beta_G' \left(H_I - H_{GH} \right)$$

In addition, the coefficient β_G' is:

$$\beta_G' = \frac{\beta_G \, P \, M_V}{R \, T}$$

The variation dH_{GH} in enthalpy of the humid gas for the interfacial area dA is then given by:

$$W_G \, \frac{dH_{GH}}{dA} = \beta_G' \left(H_I - H_{GH} \right)$$

By integrating, we find the total heat exchange surface area A necessary:

$$A = \frac{W_G}{\beta_G'} \int_{E_2}^{E_1} \frac{dH_{GH}}{H_I - H_{GH}}$$

We can divide the arc E_1E_2 in Figure 1.3 into n intervals. At each point E_i, we would know H_{Ii} and t_{Ii}. From this, we deduce H_{GH} and t_E by the intersection of the operating line with the correspondence line passing through H_{Ii} and t_{Ii}.

The interfacial surface area A is then:

$$A = \frac{W_G}{\beta_G'} \sum_1^n \left[\frac{\Delta H_{GHi}}{H_{Ii} - H_{GHi}} \right]$$

where:

$$\Delta H_{GHi} = \frac{1}{2} \left(H_{GH,i+1} - H_{GH,i-1} \right)$$

NOTE.–

There are cross-current cooling towers in which liquid air circulates horizontally. In such cases, we need to bring in the correction coefficient F, which is that used for heat exchangers. We then have:

$$A_{cross} = \frac{A_{parallel}}{F}$$

F can be calculated using the relations and curves found by [BOW 40].

1.4.3. *Cross-current tower*

Certain authors introduce the overall coefficient of enthalpy transfer K'. To do so, they simply write:

$$\beta'_G \left(H_{GHI} - H_{GH} \right) = K' \left(H^* - H_{GH} \right) = K' \left(H^* - H_{GHI} \right) + K' \left(H_{GHI} - H_{GH} \right)$$

(where H* = H* (t_E), equation of the equilibrium curve).

Divide the first and third terms in these two equations by β'_G (H_{GHI} − H_{GH}) K':

$$\frac{1}{K'} = \frac{1}{\beta'_G} + \frac{\left(H^* - H_{GHI} \right)}{\beta'_G \left(H_{GHI} - H_{GH} \right)}$$

However:

$$\beta'_G \left(H_{GHI} - H_{GH} \right) = \alpha_L \left(t_E - t_I \right) > 0$$

Finally:

$$\frac{1}{K'} = \frac{1}{\beta'_G} + \frac{\left(H^* - H_{GHI} \right)}{\alpha_L \left(t_E - t_I \right)} = \frac{1}{\beta'_G} + \frac{C}{\alpha_L}$$

The coefficient C is the slope of the equilibrium curve (E).

The transfer surface is then, with a cross-current:

$$A = \frac{W_G}{K'F} \int_2^1 \frac{dH}{H^* - H_{GH}}$$

F is the correction coefficient found by [BOW 40].

1.4.4. *White plume of air exiting the tower (in winter)*

If the plume of white vapor blows across a road or an airport runway, for instance, serious accidents could occur.

The outgoing air, which is very humid, mixes with the very cold ambient air. In the Mollier diagram, the gas resulting from the mixture is the barycenter of the images of the two gases making up the mixture. If that barycenter is on the side oversaturated in relation to the saturation curve, we see a fog forming. In order to avoid this, we simply need to warm up the air exiting the tower. Indeed, the thermal power needed will be much less than the thermal power exchanged in the tower, and the expense will be moderate (though not very eco-friendly). The two possible means of heating are:

– a gas burner;

– a finned-tube heat exchanger.

[CAM 76] discusses an example of each of these ways of working.

1.4.5. *Natural-draft tower*

We shall suppose that the thermal study of the tower has been carried out for various values of the flowrate W_G of air counted dry. Thus, we know the value of t_{GH} and Y_{GH} as a function of the height h in the tower for each value of W_G.

The density of the humid air can be deduced from this:

$$\rho_{GH}\left(h, W_G\right) = \rho_{GH}\left(t_{GH}\left(h\right), Y_{GH}\left(h\right), W_G\right)$$

The static air pressure at the base of the tower is then:

$$P_{Ba}\left(W_G\right) == g\sum_1^n \rho_{GH} dh$$

If the external pressure of the atmosphere is P_{Atm}, then the driving pressure is:

$$P_{Mo}\left(W_G\right) = P_{Ba}\left(W_G\right) - P_{Atm} = P_{Ba}\left(W_G\right) - P_{Atm} gh$$

The pressure drop in the tower due to friction is:

$$\Delta P_F(W) = \zeta \sum_1^n \rho_{GH} V_{GH}^2 dh$$

V_{GH} is the velocity of the humid gas in the tower and ζ is the pressure-drop coefficient, which depends on the geometry of the tower.

$$V_{GH}(h, W_G) = \frac{W_G(1 + Y(W_G, h))}{\rho_{GH}(h, W_G)A}$$

A: area section of the tower available to the air: m².

The driving pressure must balance out the forces of friction.

$$\rho_{Mo}(W_G) = \Delta P_F(W_G)$$

This equation sets the value of the air flowrate counted dry W_G, which must pass through the tower in order, by natural draft, to obtain a given degree of cooling for the water.

Furger [FUR 68] proposed a different, less-explicit calculation method.

2

Adsorption and
Ion-exchange Chromatography

2.1. General

2.1.1. *Adsorption of a gaseous impurity*

The principle of adsorption of gaseous impurities is to run the gaseous mixture to be purified through a column of adsorbent. The impurity is fixed to the particles of adsorbent. The second step is to regenerate the adsorbent by subjecting it to the passage of a gas which desorbs and entrains the impurity.

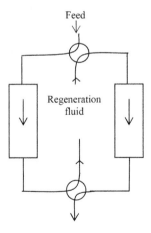

Figure 2.1. *Adsorption and regeneration*
in columns with four-way valves

Desorption can be achieved either by lowering the pressure or by increasing the temperature.

The transition between adsorption and regeneration is achieved by rotating the two quarter-turn valves by 90 degrees.

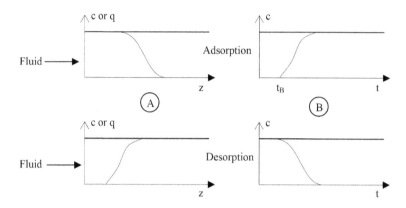

Figure 2.2. *Adsorption and desorption*

In Figure 2.2, we see:

– in part A on the left, the profiles of the concentrations in the column;

– in part B on the right, the evolution over time of the concentration in the fluid output from the column.

In Figure 2.2(B), pertaining to adsorption, starting at time t_B, the output concentration leaves the value near to zero which was specified. Thus, the specification is no longer satisfied. We say that it is broken. *The time t_B is the breaking time*. The curves in B are breakthrough curves.

2.1.2. *The different ways of working in chromatography [TIS 43]*

There are three modes of operation in chromatography.

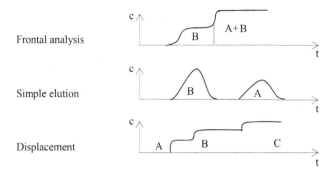

Figure 2.3. *Elution curves for solutes*

2.1.3. *Frontal analysis*

In frontal analysis (Figure 2.3), the region between the first and second fronts pertains to the solute that is least kept in the pure state. All the other regions pertain to mixtures.

Suppose the column is imbibed with pure solvent. We then introduce the solution of A, whose concentration is c_A.

By the piston effect, the solution pushes the pure solvent out of the column, which corresponds to the volume $v_c\varepsilon$. Then, the solid in the column becomes charged in the solute A until it reaches saturation. At that point:

$$v_c\left(1-\varepsilon\right)\rho_A = v_r c_A$$

The v_r is the so-called "delay" volume. During the saturation of the column, the solution has been stripped of the solute A, and the solvent comes out in the pure state. The corresponding volume of liquid is v_r.

The volume of pure solvent recovered at the output is therefore:

$$V_A = \varepsilon v_c + v_r$$

Then, the solution passes through the column without any modification. Note that the equilibrium coefficient is:

$$K_A = \frac{q_A}{c_A} = \frac{v_r}{\left(1-\varepsilon\right)v_c}$$

Thus, we have the expression of the delay volume $v_r = K_A (1 - \varepsilon) v_c$.

If the isotherm is not linear and if it is favorable, then the ratio $K_A = q_A/c_A$ will decrease when c_A increases, and v_r will decrease too.

Let us now examine the case where two solutes are present and where A is less adsorbed than B – i.e.:

$$K_A < K_B \quad \text{and} \quad v_A < v_B$$

When A and B have been adsorbed, a volume $(v_B - v_A)$ at the observed concentration c_{Aobs} has come out of the column. In the column remains $K'_A (1-\varepsilon) v_c c_A$ in the solid and finally $\varepsilon\, v_c\, c_A$ in the liquid volume of the column. These three volumes are, together, equal to the total volume injected:

$$v_B c_A = (v_B - v_A) c_{Aobs} + K'_A (1-\varepsilon) v_c c_A + \varepsilon\, v_c c_A$$

where c_{Aobs} is the concentration of A in the effluent during the fixation of B, and K'_A is the equilibrium coefficient of A in the presence of B. We deduce:

$$\frac{c_{Aobs}}{c_A} = \frac{v_B - K'_A (1-\varepsilon) v_c - \varepsilon v_c}{v_B - v_A}$$

It is only when $K_A = K'_A$ that $c_{Aobs} = c_A$, meaning that the concentration c_A of the feed is equal to the concentration c_{Aobs} of the volume $v_B - v_A$. Otherwise, all the terms involved in that relation can be found directly by measuring, except for K'_A, which we can then obtain by using the above relation. Remember that K'_A is the equilibrium coefficient of A in the presence of B.

By using "primes" to indicate the equilibrium coefficients K'_A and K'_B of each solute in the presence of the other, we can write:

$$\frac{c_{Aobs}}{c_A} = \frac{K'_B - K'_A}{K'_B - K_{Aobs}}$$

K_{Aobs} is the equilibrium coefficient of A alone present in the solution.

Often, the affinities for the solid are very different, and this analysis method loses accuracy.

2.1.4. Development by elution (with the initial solvent)

After having fixed the solutes in the column, we perform elution with the same solvent as that of the solution A and B, i.e. the initial solvent.

In the filtrate, we obtain "bands" of concentration which are essentially rectangular if the isotherms A and B are linear and independent of one another. Otherwise, we obtain two concentration profiles in the form of distinct, asymmetrical bell curves, and these curves become flatter as the height of the column increases.

When adsorption is weak, it is advisable to use the initial solvent, because then we are near to the linear case.

The *resolution* between two peaks (or two bands) obtained by elution is expressed by:

$$R = \frac{t_{Ri} - t_{Rj}}{\left(\sigma_i + \sigma_j\right)}$$

The standard deviations σ_i and σ_j are obtained as shown in Figure 2.5(A). The property which takes care of the quality of the resolution is the selectivity.

$$s_{ij} = \frac{q_i c_j}{c_i q_j}$$

2.1.5. Development by displacement ("pushing")

If the development is carried out with a strongly-adsorbed solvent different to the initial solvent, the solute moves in the column *in front of* the solvent, which pushes it forward. Thus, if there are multiple solutes adsorbed in the column, we see the appearance of a series of zones, clearly separated from one another (without mutual mixing). Each zone corresponds to a solute, and its concentration no longer depends on the length of the column. The concentration of each zone is characteristic of each solute. The drawback of this approach is that each zone is in direct contact with those surrounding it, and there are no separations with pure solvent. Therefore, we need to introduce solutes with intermediary affinities, to separate the zones we wish to isolate. The thickness of a zone is proportional to the quantity of solute introduced.

The velocity U of the train of "bands" is determined by the pusher (see section 2.5.2).

$$U = \frac{u}{1 + \dfrac{1-\varepsilon}{\varepsilon} \dfrac{dq_P}{dc_P}}$$

q_P and c_P: concentration of the pusher in the solid and in the liquid: $kmol.m^{-3}$.

In order for the solutes to move at the same velocity, we need to be on their isotherm at a concentration such that:

$$\frac{dq_i}{dc_i} = \frac{dq_P}{dc_P} = const. \qquad\qquad \forall i$$

If that equality is not possible for a solute k, it is because it is very poorly adsorbable and because we have:

$$\frac{dq_i}{dc_k} < \frac{dq_P}{dc_P} \qquad\qquad \text{whatever the value of } c_K$$

An independent band will then progress at the head of the train.

Using the "pusher" method, thus, we can separate out solutes which are present at a very low quantity in the feed.

Consider two solutes such that A is adsorbed less than B.

If the two solutes A and B have a Langmuirian isotherm, at the end of the band of A, the concentrations c_A and q_A decrease, and the equilibrium coefficient $(dq/dc)_A$ increases, so that gradient may become equal to the value $(dq/dc)_B$ and, therefore, we may see a mixture of bands A and B. This phenomenon limits the usefulness of the pusher method. One possible solution is, in advance, to adsorb the product A to the solid in a very small quantity before carrying out the adsorption-pushing operation.

2.2. Modes of feeding an adsorption column

2.2.1. *Rectangular injection – brief injection*

Over a time period t_p, we introduce the volume V_0 of fluid at a constant concentration c_0 (or, better put, at a constant composition).

Now consider the Dirac distribution (generally known as brief injection). To do so, consider a number ε as small as possible, and a volume of fluid V_0. We can choose a volume V such that:

$$|V - V_0| < \varepsilon$$

According to Dirac, we can choose a concentration c_0 such that for $c > c_0$, we have:

$$c \times |V - V_0| = M$$

M is expressed in kg or kmol. We then write:

$$M = M \times \delta(V_0)$$

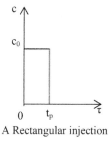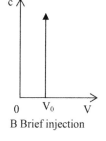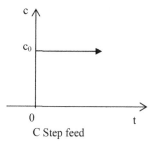

A Rectangular injection   B Brief injection   C Step feed

Figure 2.4. *Feeding of a column*

Indeed, $\delta(V_0)$ has a modulus equal to 1. This is the Dirac distribution at the point $V = V_0$ (see Figure 2.4(B)). The Dirac distribution is the idealization of a feed of low mass and quasi-instantaneous (brief injection).

2.2.2. Step feed

We may decide to feed an adsorption column from time $t = 0$ onwards, with a constant concentration of the fluid c_0, and even more generally with a constant composition. This is what is known as a step concentration profile (see Figure 2.4(C)).

2.3. Transfer parameters

2.3.1. *Axial dispersivity of the fluid phase*

The coefficient of dispersion is called dispersivity here, because it is measured in the same units as diffusivity: namely $m^2.s^{-1}$.

The dispersivity D_a can be obtained for gases and liquids by Gunn's expression [GUN 87].

$$\frac{D_a}{U_1 d} = \frac{ReSc}{4\alpha_1^2(1-\varepsilon)}(1-p)^a + \frac{Re^2Sc^2}{16\alpha_1^4(1-\varepsilon)^2}p(1-p)^3\left[\exp\left(\frac{-4(1-\varepsilon)\alpha_1^2}{p(1-p)ReSc}\right)-1\right] + \frac{\varepsilon}{\tau ReSc}$$

α_1: first root of the zero-order Bessel function $J_0(x)$ for the first species

$$\alpha_1 = 2.4048$$

ε: porosity (empty fraction in volume) of the bed;

Re: Reynolds number

$$Re = \frac{du\rho}{\mu}$$

u: interstitial velocity of the fluid: m.s^{-1}

$$u = \frac{U}{\varepsilon}$$

U: velocity of the fluid in an empty bed: m.s^{-1};

d: diameter of the particles: m;

ρ:density of the fluid: kg.m^{-3};

Sc: Schmidt number

$$Sc = \frac{\mu}{\rho D_m}$$

μ: viscosity of the fluid: Pa.s;

D_m: diffusivity of the solute: m^2.s^{-1};

p: probability for an axial displacement given by Table 2.1.

The probability p depends, amongst other things, on the shape of the particles.

Particle shape	T	P
Spheres	1.4	0.17 + 0.33 exp (-24/Re)
Cylinders	1.93	0.17 + 0.29 exp (-24/Re)

Table 2.1. *Probability for an axial displacement*

τ: tortuosity of the bed.

If the cylinder is not full but hollow, the coefficient 0.29 is replaced by the coefficient 0.20, and the tortuosity is 1.8. However, in terms of adsorption and chromatography, the particles tend to be spheroidal.

EXAMPLE 2.1.–

The fluid is a liquid progressing through a porous medium composed of spheroidal particles.

$$\varepsilon = 0.4 \qquad\qquad U = 0.01\,\text{m.s}^{-1} \qquad\qquad D_m = 10^{-9}\,\text{m}^2.\text{s}^{-1}$$

$$d = 5.10^{-5}\,\text{m} \qquad\qquad \rho = 1000\;\text{kg.m}^{-3} \qquad\qquad \mu = 10^{-3}\;\text{Pa.s}$$

$$Sc = \frac{10^{-3}}{1000 \times 10^{-9}} = 1000 \qquad Re = \frac{0.01 \times 5 \times 10^{-5} \times 1000}{10^{-9}} = 0.5$$

$$p = 0.17 + 0.33\exp\left(-24/0.5\right)$$

$$p = 0.17$$

$$\frac{D_a \times 0.4}{0.01 \times 5.10^{-5}} = \frac{0.5 \times 1000}{4 \times 2.4048^2 \times 0.6}\left(1-0.17\right)^2 + \frac{0.5^2 \times 1000^2}{16 \times 2.4048^4 \times 0.36}\times$$

$$0.17 \times 0.83^3\left[\exp\left(-\frac{4 \times 0.6 \times 2.4048^2}{0.17 \times 0.83 \times 500}\right)-1\right] + \frac{0.4}{1.4 \times 500}$$

$$\frac{D_a \times 0.4}{0.01 \times 5.10^{-5}} = 24.8 + 1297.8 \times 0.0972 \times \left(-0.173\right) + 0.00057$$

$$D_a = \frac{2.977 \times 0.01 \times 5.10^{-5}}{0.4}$$

$$D_a = 3.72.10^{-6}\,\text{m}^2.\text{s}^{-1}$$

2.3.2. Transfer coefficient for fluid–solid material transfer

[NAK 58] put forward the following correlation for the transfer coefficient k_f.

$$k_f = \frac{D}{d_p}\left(2 + 1,45\,Re^{1/2}\,Sc^{1/3}\right) \qquad\qquad (\text{m.s}^{-1})$$

D: diffusivity of the solute transferred: $\text{m}^2.\text{s}^{-1}$;

d_p: diameter of the particles: m;

Re: Reynolds number;

Sc: Schmidt number;

U: velocity of the fluid in an empty bed: $m.s^{-1}$;

μ: viscosity of the fluid: Pa.s;

ρ: density of the fluid: $kg.m^{-3}$.

EXAMPLE 2.2.–

		Liquid	Gas
D		2×10^{-9} $d_p = 5 \times 10^{-5}$ m	1.7×10^{-5}
μ		10^{-3}	16×10^{-6}
U		0.001	0.1
ρ		1000	1.3
$Re = \dfrac{Ud_p\rho}{\mu}$		0.05	0.4
$Sc = \dfrac{\mu}{\rho D}$		500	0.72
β		$\dfrac{2.10^{-9}}{5.10^{-5}}$ $\left(2+1.45\times0.05^{0.5}\times500^{0.33}\right)$	$\dfrac{1.7.10^{-5}}{5.10^{-5}}$ $\left(2+1.45\times0.4^{0.5}\times0.72^{0.33}\right)$
		$\beta = 4.10^{-5}\,m.s^{-1}$	$\beta = 0.95\;m.s^{-1}$

NOTE.–

[WIL 66] proposed a method for calculating the transfer coefficient across the liquid film, but Guiochon *et al.* [GUI 06] give that relation in the following form:

$$0.0015 < Re < 55 \qquad Sh = \frac{1.09}{\varepsilon} Re^{0.33} Sc^{0.33}$$

$$55 < Re < 1050 \qquad Sh = \frac{0.25}{\varepsilon} Re^{0.69} Sc^{0.33}$$

Sh is the Sherwood number

$$Sh = \frac{k_f d_p}{D}$$

k_f: material transfer coefficient: m.s^{-1};

d_p: diameter of the particles: m;

D: diffusivity in the liquid: m^2.s^{-1}.

2.3.3. (Balance) equation for the fluid

We have two equations: one for the fluid and the other for the solid particles. Let us begin with the fluid.

The concentration c of the solute in the fluid varies over time with the velocity dc/dt in each segment of the column, of thickness Δz. This variation is can be attributed to three phenomena:

– convection:

$$\varepsilon \left[uc - u \left(c + \frac{\partial c}{\partial z} \Delta z \right) \right] = -\varepsilon u \frac{\partial c}{\partial z} \Delta z$$

– dispersion by diffusion (Fick's law)

$$\varepsilon \left[-D_a \frac{\partial c}{\partial z} + D_a \frac{\partial c}{\partial z} + D_a \frac{\partial^2 c}{\partial z^2} \Delta z \right] = \varepsilon D_a \frac{\partial^2 c}{\partial z^2} \Delta z$$

For these equations, the notations are as follows:

ε: fraction of the volume of the column occupied by the fluid (porosity);

u_0: velocity of the fluid in an empty bed: m.s^{-1};

c: concentration of the solute in the fluid: kmol.m^{-3} or kg.m^{-3};

z: length along the column: m;

The concentration expressed per m^3 of the column is εc;

D_a: axial dispersion coefficient along the column. We can call that coefficient the *dispersivity*, by analogy to the diffusivity which is measured in the same units: m^2.s^{-1};

– exchange of solute with the solid phase. The exchange surface expressed in relation to the volume of the column is:

$$(1-\varepsilon)\frac{6}{d_p}$$

Hence, the exchange of solute in relation to the volume of the column is:

$$k(1-\varepsilon)\frac{6}{d_p}(c*-c)$$

d_p: diameter of the particles: m;

k: transfer coefficient: m.s^{-1};

c*: concentration at equilibrium with the surface of the solid.

Finally, the equation is written:

$$\frac{\partial c}{\partial t} = D_a\frac{\partial^2 c}{\partial z^2} - u\frac{\partial c}{\partial z} + k\frac{(1-\varepsilon)}{\varepsilon}\frac{6}{d_p}(c*-c) \qquad [2.1]$$

For the calculations, it is necessary to use dimensionless values i.e.:

$$\tau=\frac{t}{Lu} \qquad x=\frac{z}{L} \qquad Bi=\frac{kR}{D_s} \qquad Pé_f=\frac{u_0L}{D_a} \qquad Pé_s=\frac{u_0R}{D_s}$$

R is the radius of the particles, supposed to be spherical, and is therefore equal to $d_p/2$. The numbers Bi and Pé$_f$ are the Biot number and the Péclet number. L is the length of the column. D_s is the diffusivity in the solid.

The equation for the fluid becomes:

$$\frac{\partial c}{\partial \tau} = \frac{1}{P\acute{e}_f} \frac{\partial^2 c}{\partial x^2} - \frac{\partial c}{\partial x} - \frac{L3(1-\varepsilon)Bi(c^*-c)}{R} \frac{1}{\varepsilon} \frac{1}{P\acute{e}_s}$$

When several solutes are involved, the solute with index j assigns that index to the following properties:

$$k_j, D_{sj}, Bi_j, P\acute{e}_{fj}, P\acute{e}_{fj}$$

The index j ranges from 1 to k (which, in this instance, is the number of solutes).

NOTE.–

For diffusion inside the solid particles, the differential equation is given in section 3.5 of [DUR 16].

2.3.4. *Initial conditions and boundary conditions*

1) Fluid

At the input to the column (x = 0)

$$\frac{\partial c}{\partial x} = Pe_f (c_0 - c_F) \qquad\qquad x = 0$$

c_F: concentration of the feed;

c_0: concentration at the input.

At the output from the column:

$$\frac{\partial c}{\partial x} = 0 \qquad\qquad x = 1$$

The initial condition is expressed by the concentration profile of the fluid along the column

$$c = c(0, x) \qquad\qquad \tau=0$$

2) Particles

At the center of a spherical particle, symmetry dictates that:

$$\frac{\partial c_s}{\partial \rho} = 0 \qquad \rho = 0$$

At the surface of the particle, the exchange equation is expressed by:

$$\left. \frac{\partial c_s}{\partial \rho} \right|_{\rho = 1} = Bi \left(c^* - c \right) \qquad \rho = 1$$

The initial concentration in the particles is supposed to be uniform in each particle, but that concentration in each particle depends on the abscissa x and on the initial concentration of the fluid.

$$c_s = c_s \left(0, x \right) \text{ for } \tau = 0$$

2.4. Height equivalent to a theoretical plate

2.4.1. Graphical determination of t_R, the standard deviation σ and the parameter k'

Let us examine the instantaneous concentration c at the output from the column as a function of time. The band corresponding to the solute shows a maximum, whose abscissa value is precisely equal to the retention time t_R of the solute (see Figure 2.5(A)).

If we now take, as a function of time, the concentrations at the output of the column fed by a step function, we obtain an S curve which can be assimilated to an error function which is the integral of the Gaussian bell curve. Let us mark the abscissas of the corresponding points on the error function at 15.9% and 84.1%. The distance between these two points is precisely equal to twice the standard deviation σ of the bell curve (see Figure 2.5(B)).

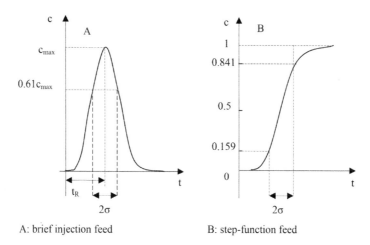

Figure 2.5. *Concentration at the output*

On condition that the column is the same in both cases, another way of measuring the difference 2σ is to plot (in Figure 2.5(A)) the horizontal with ordinate $0.61c_{max}$. This line intersects the "bell curve" at two points, whose abscissa values are the distance 2σ apart.

We know that:

$$t_R = \left(1+k'\right)t_0 = \left(1+k'\right)\frac{L}{u}$$

Hence we have the value of k' (the capacity parameter). Refer to section 2.4.2.

u: intergranular velocity of the fluid: $m.s^{-1}$;

t_0: residence time of the fluid: s.

2.4.2. Concept of capacity parameter

The capacity parameter at a given place in the column is:

$$k' = \frac{1-\varepsilon}{\varepsilon}\frac{dq}{dc}$$

If we look at the case of linear adsorption, i.e.:

$$\frac{dq}{dc} = \text{const.} = a$$

then the capacity parameter is the ratio:

$$k_0' = \frac{\text{mass of solute in the solid}}{\text{mass of solute in the fluid}}$$

We often see the situation approach linear chromatography if the concentration of the sample introduced into the column is near to zero. In this case, k' takes the value k_0':

$$\frac{dq}{dc} \# \text{const. if } c < \varepsilon \text{ (small quantities)}$$

The residence time of the fluid phase in the column is:

$$t_0 = \frac{L}{u} = \frac{L\varepsilon}{u_0}$$

u_0: velocity of the fluid in an empty bed: $m.s^{-1}$;

u: interstitial velocity of the fluid: $m.s^{-1}$.

The residence time t_0 is all the greater when the adsorption of the solute practically does not vary the fluid volume and thus the velocity u.

The time that the solute spends in the solid phase is:

$$\frac{L}{u}k' = t_0 k'$$

The overall retention time of the solute in the column is:

$$t_R = t_0 + t_0 k' = (1 + k')t_0$$

We can see that k' is a relative retention:

$$k' = \frac{t_R - t_0}{t_0}$$

The fraction of the time that a molecule of solute spends in the mobile phase is $1/(1+k')$.

The velocity of a molecule of solute is the mean between the velocity u and the velocity zero:

$$\frac{u \times 1}{1 + k'} + \frac{0 \times k'}{1 + k'} = \frac{u}{1 + k'}$$ (see section 2.5.2)

The load rate of a column is the ratio:

$$\frac{\text{adsorbed solute}}{\text{capacity of the column at saturation for that solute}}$$

More specifically, if the isotherm is Langmuirian:

$$q = \frac{ac}{1 + bc}$$

At saturation:

$$q_\infty = \frac{a}{b}$$

The capacity parameter is:

$$k' = \frac{a}{(1 + bc)^2} = \frac{k'_0}{(1 + bc)^2}$$

2.4.3. *Height equivalent to a theoretical plate [HOR 78]*

When a trace of a marker is introduced into a column in the form of a brief injection, we obtain at the output, a bell curve for the concentration of the marker. The height of a theoretical plate is then obtained by the following relation on the basis of the standard deviation of that curve.

$$H = \left(\frac{\sigma}{t_0 k_0'}\right) Z$$

σ: standard deviation of the bell curve: s;

t_0: transit time of the fluid mixture: s

$$t_0 = \frac{Z}{u}$$

Z: height of the column: m

$$k_0' = \frac{t_R - t_0}{t_0} = Fa \quad ; \quad t_R = \frac{Z}{u}\left[1 + F\frac{dq}{dc}\Big|_{c=0}\right]$$

k_0': partition coefficient for the marker, which is Fa;

u: interstitial velocity of the fluid: m.s^{-1};

t_R: marker retention time.

The reduced plate height is:

$$h = H / d_p$$

d_p: diameter of the particles: m.

[HOV 78] performed detailed calculations, and obtained the following expression:

$$h \# \frac{A}{v} + B\,v^{1/3} + D\,v \quad \text{where} \quad v = \frac{u d_p}{D}$$

u: interstitial velocity of the fluid: m.s^{-1};

D: diffusivity of the marker in the fluid.

Unfortunately, in evaluating the coefficients A, B and D, we must draw on experience.

2.4.4. *Height equivalent to a theoretical plate [KNO 77]*

This concept is inspired by liquid–gas exchange in a packed column. However, this analogy is misleading if we are looking at a chromatography column.

Indeed, for a feed in the form of a Dirac pulse, we know that the concentration profile of the liquid at the output is a bell curve whose standard deviation σ is given by the following relation, which is the definition of the TPEH.

$$\sigma^2 = HZ$$

σ^2: variance of the concentration: m^2;

Z: height of the column: m;

H: HETP (Height Equivalent to a Theoretical Plate): m.

The existence of this type of concentration profile is due to the conjunction of three random phenomena which can each be characterized by a variance. Thus, the variance σ^2 of the overall phenomenon is the sum of these three variances.

Thus, [KNO 77] defines the reduced TPEH as the sum of three terms which we shall simplify in order to facilitate a comfortable calculation.

$$h = \frac{H}{d_p} = \frac{\sigma^2}{Zd_p} = \left(\frac{ud_p}{D_m}\right)^{1/3} + \frac{2D_a\left(1+k'\right)}{ud_p} + \frac{ud_p}{120D_m}$$

u: interstitial velocity of the fluid: $m.s^{-1}$;

d_p: diameter of the particles: m;

D_m: diffusivity of the solute in the fluid: $m^2.s^{-1}$;

k': capacity parameter of the column for the solute (see section 2.4.2);

D_a: dispersivity: $m^2.s^{-1}$.

EXAMPLE 2.3.–

$$Z = 0.1 \text{ m} \qquad u = 0.01 \text{ m.s}^{-1} \qquad d_p = 2 \times 10^{-6} \text{ m}$$
$$D_a = 0.1 \times 10^{-6} \quad k' = 1.2 \qquad D_m = 1.5 \times 10^{-9} \text{ m}^2.\text{s}^{-1}$$

$$h = \left(\frac{0.01 \times 2.10^{-6}}{1.5.10^{-9}} \right)^{1/3} + \frac{2 \times 0.1.10^{-6} \times (2.2)}{0.01 \times 2.10^{-6}} + \frac{0.01 \times 2.10^{-6}}{120 \times 1.5 - 10^{-9}}$$

$$h = 2.37 + 22 + 0.16 = 24.53$$

$$N = \frac{Z}{H} = \frac{Z}{hd_p} = \frac{0.1}{24.53 \times 2 \times 10^{-6}} = 2038 \text{ theoretical plates}$$

2.4.5. Glueckauf's calculation [GLU 55] for the number of theoretical plates

The aim is to determine the evolution of the concentration of the liquid of elution of the column where the initial concentration is uniform and equal to c_0. The liquid with which we perform the elution is pure solvent.

Glueckauf's equation (4) is written:

$$\left(\frac{\partial c}{\partial x} \right)_v + a \left(\frac{\partial c}{\partial v} \right)_x = \frac{\Delta x}{2} \left(\frac{\partial^2 c}{\partial x^2} \right)_v$$

Let us perform the following substitutions:

Replace	v	a	x
with	$V - \varepsilon v$	$\dfrac{a(1-\varepsilon)}{\varepsilon}$	εv

We then find the balance equation:

$$\left(\frac{\partial c}{\partial (\varepsilon v)} \right)_{V-\varepsilon v} + \frac{a(1-\varepsilon)}{\varepsilon} \left(\frac{\partial c}{\partial (V-\varepsilon v)} \right)_{\varepsilon v} = \frac{\Delta(\varepsilon v)}{2} \left(\frac{\partial^2 c}{\partial (\varepsilon v)^2} \right)_{V-\varepsilon v} \qquad [2.2]$$

Glueckauf's solution is given by his equation 14 where, according to him, we can overlook the second term. He then transforms that equation, using (with our notations):

$$\bar{V} - \varepsilon v_c = a(1-\varepsilon)v_c$$

Indeed, at the output from the column, $v = v_c$, the volume of the column. In addition, $\bar{V}$ corresponds to the volume of elution of the center of the breakage curve (see section 2.1.1).

Glueckauf finally obtained his equation 16 for the concentration of the eluate at the output from the column:

$$\frac{c}{c_0} = \frac{1}{2} - \text{erf}(y)$$

where:

$$y = \frac{\sqrt{N}\left(\bar{V} - V\right)}{\sqrt{\left(\bar{V} - \varepsilon v_c\right)\left(V' - \varepsilon v_c\right)}}$$

N is the number of theoretical plates $v/\Delta v$.

If we set $y = 1$, we can deduce the corresponding value V' of V and, for this value:

$$\frac{c'}{c_0} = 1 - \text{erf}(1) = 0,1587$$

From this, we deduce the number of theoretical plates in the column:

$$N = \frac{\left(\bar{V} - \varepsilon v_c\right)\left(V' - \varepsilon v_c\right)}{\left(\bar{V} - V\right)^2} \quad \text{with } V' < \bar{V}$$

On anamorphozed paper (Figure 1 in Glueckauf's publication), the function $c/c_0 = f(y)$ is represented by a straight line whose parameter is N.

In [GLU 55] Figure 1, we write the point corresponding to $(V - \varepsilon v_c)/(V' - \varepsilon v_c)$ and to $c/c_0 = 0.1587$. On the line passing through that point, we read the value of N.

2.5. Ideal system and material balance

An ideal system is one in which the *height equivalent to a theoretical plate is null*. The result of this is that:

– the axial dispersivity is null;

– the transfer coefficient between fluid and particles is infinite, which means that local equilibrium is achieved between fluid and particles.

Furthermore, in an ideal system, the isotherm is linear.

We shall use the following notations:

ε: empty fraction of the bed of divided solid;

u: interstitial velocity of the fluid: $m.s^{-1}$;

c_i: concentration of the component i in the fluid: $kg.m^{-3}$ or $kmol.m^{-3}$;

q_i: concentration of the component i in the solid: $kg.m^{-3}$ or $kmol.m^{-3}$;

z: distance along the axis of the column counted positively in the direction of progression of the fluid.

The equation of the isotherm is:

$$q_i = f\left(c_1, c_2, ..., c_i, ..., c_n\right)$$

In order to establish the *material balance equation*, let us write that the rate of enrichment over time of a segment of the column of thickness Δz is equal to the difference between what comes into the segment and what goes out.

$$\left[\varepsilon\frac{\partial c_i}{\partial \tau} + (1-\varepsilon)\frac{\partial q_i}{\partial \tau}\right]\Delta z = \left[uc_i - \left(uc_i + \frac{\partial(uc_i)}{\partial z}\right)\right]\varepsilon\Delta z$$

$$\frac{\partial c_i}{\partial t} + \left(\frac{1-\varepsilon}{\varepsilon}\right)\frac{\partial q_i}{\partial t} + \frac{\partial(uc_i)}{\partial z} = 0$$

We can write the following, supposing that, *at all points, equilibrium exists between the fluid and the solid*:

$$\frac{\partial q_i}{\partial c_i} = f'(c_1,...,c_i,...c_n)$$

Finally, the material balance equation is written as follows for an ideal system:

$$\left[\frac{(1-\varepsilon)}{\varepsilon}f'+1\right]\frac{\partial c_i}{\partial t} + u\frac{\partial c}{\partial z} = 0$$

Still in the ideal system, we suppose that the relation linking q_i and c_i is linear:

$$q_i = r_i c_i \; ; \; \frac{dq_i}{dc_i} = r_i = \text{const.}$$

This greatly simplifies the function f. The equilibrium of a component does not depend on the other components.

2.5.1. *Quasi-ideal system*

In a quasi-ideal system, the height HETP is always equal to zero. On the other hand, the equation of the equilibrium isotherm is no longer linear. The following expressions are *the most widely used*.

$$q_i = \frac{ac_i}{1+bc_i} \qquad \text{(Langmuir)}$$

$$q_i = ac_i^{1/n} \qquad \text{(Freundlich)}$$

In the system, it is accepted that there is still local equilibrium (i.e. everywhere) between the fluid and the solid.

2.5.2. *Constant concentration lines in the fluid*

We shall set:

$$F = \frac{1 - \varepsilon}{\varepsilon}$$

The material balance becomes:

$$\frac{\partial c_i}{\partial t}(1 + Ff') + u\frac{dc_i}{dz} = 0$$

As shown in Appendix 2, for a closed-loop derivation, we can write:

$$\left.\frac{\partial c}{\partial z}\right|_t = \left.\frac{\partial t}{\partial z}\right|_c \times \left.\frac{\partial c}{\partial t}\right|_z$$

Let us eliminate $\dfrac{\partial c}{\partial z}$ between this equation and the material balance.

$$(1 + Ff')\left.\frac{\partial c}{\partial t}\right|_z = u\left.\frac{\partial t}{\partial z}\right|_c \left.\frac{\partial c}{\partial t}\right|_z$$

Let us simplify by $\left.\dfrac{\partial c}{\partial t}\right|_z$. We obtain the following for the speed of progression of a wave of constant concentration:

$$\frac{u}{1 + Ff'} = \frac{1}{\left(\partial t / \partial z\right)_c} = \left.\frac{\partial z}{\partial t}\right|_c = U \qquad \text{(Appendix 2: theorem of reciprocity)}$$

For an ideal system, the curves with c=const. are straight lines, known as characteristics (indeed, f'=const.).

For a quasi-ideal system, the characteristics are no longer rectilinear.

NOTE.–

The material balance equation is of the form:

$$\alpha \frac{\partial c}{\partial t} + \beta \frac{\partial c}{\partial z} = \gamma$$

The equation of the characteristics for such a hyperbolic equation is (see [NOU 85]):

$$\frac{dt}{\alpha} = \frac{dx}{\beta} = \frac{dc}{\gamma}$$

The first equality confirms the result found previously:

$$U = \frac{u}{1 + Ff'}$$

The characteristics corresponding to $\gamma = 0$ are $dc = 0$ and c is a constant for these characteristics.

2.5.3. *Material balance with two terms (quasi-ideal system)*

Consider the function m (x,y). The derivative $\left(\dfrac{\partial m}{\partial y}\right)_{x-y}$ corresponds to a variation in m for a variation in y.

However, we must preserve the relation:

$$x - y = \text{const., so } \Delta x = \Delta y$$

Thus, besides the variation in y, we need to take account of the variation in x. Finally:

$$\left(\frac{\partial m}{\partial y}\right)_{x-y} \Delta y = \left(\frac{\partial m}{\partial y}\right)_x \Delta y + \frac{\partial m}{\partial x} \Delta x = \left[\left(\frac{\partial m}{\partial y}\right)_x + \left(\frac{\partial m}{\partial x}\right)_y\right] \Delta y$$

$$\left(\frac{\partial m}{\partial x}\right)_{x-y} = \left(\frac{\partial m}{\partial x}\right)_x + \left(\frac{\partial m}{\partial x}\right)_y$$

We set:

$$m = c \qquad x = V \qquad y = \varepsilon v$$

We then find:

$$\left(\frac{\partial c}{\partial (\varepsilon v)} \right)_{V-\varepsilon v} = \left(\frac{\partial c}{\partial (\varepsilon v)} \right)_{V} + \left(\frac{\partial c}{\partial V} \right)_{\varepsilon v} \qquad\qquad [2.3]$$

However, the material balance is written:

$$-\frac{(1-\varepsilon)}{\varepsilon} \left(\frac{\partial q}{\partial V} \right)_{\varepsilon v} = \left(\frac{\partial c}{\partial V} \right)_{\varepsilon v} + \left(\frac{\partial c}{\partial \varepsilon v} \right)_{V} \qquad\qquad [2.4]$$

By subtracting the second equation from the first, we obtain:

$$\left(\frac{\partial c}{\partial v} \right)_{V-\varepsilon v} + \frac{(1-\varepsilon)}{\varepsilon} \left(\frac{\partial q}{\partial V} \right)_{v} = 0$$

or indeed:

$$\left(\frac{\partial c}{\partial \varepsilon v} \right)_{V-\varepsilon v} + (1-\varepsilon) \left(\frac{\partial q}{\partial (V-\varepsilon v)} \right)_{\varepsilon v} = 0 \qquad\qquad [2.5]$$

2.5.4. Sillén's law [SIL 50]

Let us set:

$$T = \frac{V - \varepsilon v}{\varepsilon v}$$

T is the crossing parameter.

or indeed:

$$\varepsilon v = \frac{V - \varepsilon v}{T} \quad \text{and} \quad (d(\varepsilon v))_{V-\varepsilon v} = -\frac{(V - \varepsilon v)}{T^2} dT$$

Consequently,

$$\left(\frac{\partial c_i}{\partial (\varepsilon v)}\right)_{V-\varepsilon v} = -\frac{T^2}{(V-\varepsilon v)}\frac{dc_i}{dT} = \frac{-T}{\varepsilon v}\frac{dc_i}{dT}$$

In addition:

$$V-\varepsilon v = \varepsilon v T \quad \text{and} \quad \left(d(V-\varepsilon v)\right)_{\varepsilon v} = \varepsilon v dT$$

Therefore:

$$\left(\frac{\partial q_i}{\partial (V-\varepsilon v)}\right)_{\varepsilon v} = \frac{1}{\varepsilon v}\frac{dq_i}{dT} = \frac{1}{\varepsilon v}\frac{dq_i}{dc_i}\frac{dc_i}{dT}$$

The balance equation then becomes:

$$-\frac{1}{\varepsilon v}\frac{dc_i}{dT}\left(\frac{dq_i}{dc_i}-T\right)=0$$

This obviously assumes that we have $c =$ const., but it is the other solution which is of interest to us:

$$\frac{dq_i}{dc_i}=T=\frac{V-\varepsilon v}{\varepsilon v}$$

We see that the crossing parameter T is the same regardless of the nature of the solute and that, consequently, the ratio dq_i / dc_i is independent of the nature of the solute.

2.5.5. *Consequence of Sillén's law*

We have seen that the equilibrium of ions with the same valences is written:

$$q=\frac{QKc}{C^*+(K-1)C}=\frac{QKx}{1+(K-1)x}\qquad\left(\text{where } x=\frac{c}{c^*}\right)$$

We set:

$$\frac{q}{Q} = y \quad \text{and} \quad K = r$$

Hence:

$$y = \frac{rx}{1 + (r-1)x}$$

According to Sillén's law:

$$T^* = \frac{dy}{dx} = \frac{r}{\left(1 + (1-r)x\right)^2}$$

Note that we can write:

$$T = \frac{dq}{dc} = \frac{Q}{C^*} \frac{dy}{dx} = \frac{Q}{C^*} T^*$$

Hence:

$$x = \frac{r - \sqrt{r/T^*}}{r - 1}$$

We must have $0 < \dfrac{r - \sqrt{r/T^*}}{r-1} < 1$ 　 We must have $0 < \dfrac{\sqrt{r/T^*} - r}{1 - r} < 1$

If r > 1　　　　　　　　　　　　　　　If r < 1

and x is an increasing function of T^* and x is a decreasing function of T^*

Using equations 9b, 18 and 38 from [WAL 45], [TON 67], in their equations 8 and 9 generalized the expression of x for a mixture of several solutes. The solutes were ions of the same valences. The authors deduce the evolution of the concentrations of the solid phase in the column.

[KLE 67] dealt with the problem of ions with different valences, again using Sillén's law.

2.6. Analytical solutions

2.6.1. *Rosen's solution for a step feed in a column [ROS 52. ROS 54]*

The author supposes that the *isotherm is linear:*

$$q = Kc$$

q: concentration in the solid: $kmol.m^{-3}$ or $kg.m^{-3}$;

c: concentration in the fluid: $kmol.m^{-3}$ or $kg.m^{-3}$;

K: partition coefficient: dimensionless.

We shall let $\bar{q}$ represent the mean concentration of a particle at level z of the column.

$$\bar{q}(x,\theta) = \frac{3}{R_p^3} \int_0^{R_p} q(r,x,\theta)\, r^2 dr$$

R_p: radius of the particle: m

$$x = \frac{z}{\varepsilon v} \qquad \theta = t - \frac{z}{v}$$

v is the interstitial velocity of the fluid, although Rosen speaks of the surface velocity (in an empty bed);

ε: porosity (void fraction) of the bed of particles.

The quantity of solute entering a particle per unit time is:

$$4\pi R_p^2 \left(c - \frac{q_s}{K} \right) = \frac{4}{3}\pi R_p^3 \frac{d\bar{q}}{dt}$$

which is written:

$$\frac{d\overline{q}}{dt} = \frac{3k_f}{R_p}\left(c - \frac{q_s}{K}\right) = \frac{1}{R_f}\left(c - \frac{q_s}{K}\right)$$

k_f: transfer coefficient on the side of the fluid: m.s^{-1};

q_s: concentration at the surface of a particle.

The group $R_f = \dfrac{R_p}{3k_f}$ is the resistance of the fluid film.

Initially, the concentration $q_0 = \overline{q}_0$ of the solid is at equilibrium with the concentration of the impregnating fluid.

Over the course of the adsorption, we can set:

$$u(x,\theta) = \frac{c(x,\theta) - (q_0 / K)}{c_0 - (q_0 / K)}$$

Here, c_0 is the concentration, supposed to be constant, of the feed liquid during the course of the adsorption.

Rosen's solution [ROS 52] is written:

$$u(x,\theta) = \frac{1}{2} + \frac{2}{\pi}\int_0^\infty e^{-\gamma x H_1(\lambda,r)} \times \sin\left[\sigma\theta\lambda^2 - \gamma x H_2(\lambda,v)\right]\frac{d\lambda}{\lambda}$$

The notations used are as follows:

$$\gamma = \frac{3D_s K}{R_p^2} \qquad \sigma = \frac{2D_s}{R_p^2} \qquad v = \gamma R_f = \frac{D_s K}{R_p k_f} \qquad \theta = t - \frac{z}{v}$$

The expression of H_1 and that of H_2, which are functions of v and λ, are given in Table I of Rosen's book [ROS 52].

In order to take account of the presence of the sine, in performing the integration, we must choose a high number of intervals (trapezoidal method) – around a thousand – but this does not pose problems, because the computers of today are far faster than those in Rosen's time.

When the column is sufficiently high, meaning that x is high, the function u tends toward a limit:

$$u = \frac{1}{2}\left[1 + \mathrm{erf}\left[\frac{\dfrac{3y}{2x} - 1}{2\sqrt{v/x}}\right]\right] \qquad \text{[ROS 54]}$$

where:

$$x = \frac{3}{\varepsilon v} \qquad y = \frac{2D_s\left(t - \dfrac{z}{v}\right)}{R_p^2}$$

[ROS 52] also treats the nullity of the resistance of the fluid film – i.e. $v = 0$ – as a limiting case.

In his Figures 2 and 3 in [ROS 54], Rosen gives lattices of curves $u = f(y/x)$ parameterized in x. For each lattice, there is a corresponding value de v/x.

2.6.2. Linear isotherm [THO 48]. Step injection

Instead of describing the exchange of material between a fluid and solid in the traditional two steps (across the liquid film surrounding the solid particles and within those particles), the author uses the language of chemical reaction, writing the following for the material transfer function:

$$\frac{dq}{dt} = f(\overline{q}, c) = k_1^* c - k_2^* q$$

Hereafter, we shall set (see balance equation [2.5])

$$k_1 = k_1^* F \qquad k_2 = k_2^* F \quad \text{where} \quad F = \frac{1 - \varepsilon}{\varepsilon}$$

Thomas uses the groups which he calls Ay and Bx.

$$Ay = \frac{k_2(V - \varepsilon v)}{\dot{V}} \qquad Bx = \frac{k_1 \varepsilon v}{\dot{V}} \qquad \text{(see equation [26.5])}$$

Let us set:

c_0: concentration of the fluid at the input to the column;

q_0: concentration of the solid at equilibrium with c_0.

For saturation, the results are:

$$\frac{c}{c_0} = e^{-(Bx+Ay)} \left[\varphi(Ay, Bx) + I_0 \left(2\sqrt{ABxy} \right) \right]$$

$$\frac{\overline{q}}{q_0} = e^{-(Bx+Ay)} \varphi(Ay, Bx)$$

I_0 (z) is the zero-order Bessel function whose serial expansion is:

$$I_0 \left(2\sqrt{uv} \right) = \sum_{m=0}^{\infty} \frac{u^m v^m}{m!m!}$$

The function φ (u,v) is defined by:

$$\varphi(u, v) = e^u \int_0^u e^{-t} I_0 \left(2\sqrt{vt} \right) dt = \sum_{0 \le n < m} \frac{u^n v^m}{m!n!}$$

For the elution, Thomas obtained:

$$\frac{c}{c_0} = e^{-(Bx+Ay)} \left[\varphi(Bx, Ay) \right]$$

$$\frac{\overline{q}}{q_0} = e^{-(Bx+Ay)} \left[\varphi(Bx, Ay) + I_0 \left(2\sqrt{AB\,xy} \right) \right]$$

q_0: initial concentration of the bed;

c_0: concentration at equilibrium with q_0.

NOTE.–

We can write the material transfer function:

$$\frac{dq}{dt} = k_1^* c - k_2^* q = k_f \left(c - c^* \right) = k_f \left(c - \frac{q}{K} \right)$$

where:

$$k_1^* = k_f \qquad k_2^* = \frac{k_f}{K}$$

K is the partition constant $K = q/c$, with the isotherm supposed to be linear.

In Thomas' approach, the concentration of solute in the solid only comes into play in the form of its mean value $\bar{q}$ over the whole volume of each particle. Therefore, $\bar{q}$, like the concentration c of the fluid, depends only on two variables t and z or, more specifically, on the variables Ay and Bx.

V: fluid volume injected into the column: m^3;

v: volume of the column between the inlet and the abscissa z: m^3;

ε: empty fraction in the solid bed: dimensionless;

$\dot{V}$: volumetric flowrate of the fluid: $m^3.s^{-1}$;

k_f: material transfer coefficient across the fluid film: $m.s^{-1}$;

K: partition coefficient (at equilibrium) between the two phases: dimensionless;

q and c: concentration in the solid and in the fluid: $kg.m^{-3}$ or $kmol.m^{-3}$.

Unlike Rosen's solution, here the mean concentration $\bar{q}$ replaces Rosen's concentration q_s at the surface of a particle.

2.6.3. *Ion exchange where the ions have the same valence* [THO 44]

$$A^{a+} + B\ Resin = B^{a+} + A\ Resin$$

$$\frac{dq}{dt} = k_1^*(Q-q)c - k_2^*(c_0 - c)q$$

Hereinafter, we shall set:

$$k_1 = k_1^*F \qquad k_2 = k_2^*F \quad \text{where} \quad F = \frac{1-\varepsilon}{\varepsilon}$$

The isotherm is given by the law of mass action:

$$\frac{k_2}{k_1} = \frac{(Q-q)c}{(c_0 - c)q} = K$$

$$Ay = \frac{k_2 c_0 (V - \varepsilon v)}{\dot{V}} \qquad\qquad Bx = \frac{k_1 Q\varepsilon v}{\dot{V}}$$

$$\alpha y = \frac{k_1 c_0 (V - \varepsilon v)}{\dot{V}} \qquad\qquad \beta x = \frac{k_2 Q\varepsilon v}{\dot{V}}$$

For example, in the case of adsorption:

$$\frac{c}{c_0} = \frac{I_0\left(2\sqrt{ABxy}\right) + \varphi(\alpha y, \beta x)}{I_0\left(2\sqrt{ABxy}\right) + \varphi(Bx, Ay) + \varphi(\alpha y, \beta x)}$$

$$\frac{q}{Q} = \frac{\varphi(\alpha y, Bx)}{I_0\left(2\sqrt{ABxy}\right) + \varphi(Bx, Ay) + \varphi(\alpha y, \beta x)}$$

V: volume fed in from time zero: m^3;

εv: volume available to the liquid in the length z of the column: n

$$v = Sz$$

S: section of the column: m²;

$\dot{V}$: flowrate of fluid fed in: m³.s⁻¹.

2.6.4. Coefficient of reaction kinetics [NEL 56]

Exchange of ions with the same valence is expressed by the relation:

$$\frac{dq}{dt} = \frac{\kappa}{Q}\left[c(Q-q) - \frac{1}{K}q(c_0 - c)\right]$$

κ: kinetic constant (pronounced "kappa"): s⁻¹;

K: equilibrium constant: dimensionless.

The constant κ is given by:

$$\frac{ab}{\kappa} = \frac{1}{k_f} + \frac{c_0}{Qk_p}$$

The transfer coefficients k_f and k_p are given in sections Chapter 3 of [DUR 16c] and Chapter 4 of this book and are measured in m.s⁻¹.

a: volumetric surface area of the solid: m².m⁻³;

b is a constant dimensionless function of r = 1/K and of the parameter ζ (pronounced "zeta").

$$\zeta = \frac{Qk_p}{c_0 k_f}$$

The authors give analytical expressions for b and, in particular, a family of curves b = f (ζ) parameterized in r.

In order to form the link with the calculation from [THO 44], we simply need to set:

$$k_1^* = \frac{\kappa}{Q} \quad \text{and} \quad k_2^* = \frac{\kappa}{QK}$$

NOTE.–

[HIE 52] put forward lattices of curves c/c_0 as a function of t/s parameters in s. Each lattice of curves corresponds to a value of the parameter r.

Type of sorption	R	t	S
Ion exchange	k_2 / k_1	$\dfrac{k_1 c_0 \left(V - \varepsilon v\right)}{\dot{V}}$	$\dfrac{\varepsilon v}{\dot{V}}$
Langmuirian	$\dfrac{k_2}{k_2 + k_1 c_0}$	$\dfrac{c_0 k_1 + k_2 \left(V - \varepsilon v\right)}{\dot{V}}$	$\dfrac{\varepsilon v}{\dot{V}}$

Table 2.2. *Parameters used by Hiester and Vermeulen*

[HIE 52] do not touch on the case of the linear isotherm.

[VER 52] precisely define the general meaning of the parameters t and s.

NOTE.–

If the flowrate of the fluid in terms of volume $\dot{V}$ tends toward zero, thermodynamic equilibrium is established between the solid and fluid. [GOL 53] studied the asymptotic behavior of the function used by [HIE 52].

$$J(x,y) = 1 - e^{-x-y} \, \varphi(x,y)$$

[GOU 65] studied the asymptotic behavior of the Bessel functions.

2.6.5. *Calculation of the constant concentration profile [COO 65]*

When the isotherm is said to be favorable, i.e. when its concave face is turned toward the axis of the concentrations in the fluid, experience of the calculations shows us that the profile of q(z) and that of c(z) become independent of the score of depth z.

The balance equation is written:

$$D_a \frac{\partial^2 C}{\partial z^2}\bigg|_t = v \frac{\partial c}{\partial z}\bigg|_t + \frac{\partial c}{\partial t}\bigg|_z \frac{1-\varepsilon}{\varepsilon} \frac{\partial q}{\partial t}\bigg|_z$$

We set:

$$z^* = z - Ut \quad \text{and} \quad U = \frac{u}{1 + \frac{(1-\varepsilon)}{\varepsilon} \frac{\Delta q}{\Delta c}}$$

If t = const. then dz = dz*

If z = const. then $dt = -\dfrac{dz^*}{U}$

Thus:

$$v \frac{\partial c}{\partial z}\bigg|_t + \frac{\partial c}{\partial t}\bigg|_z = v \frac{dc}{dz^*} - U \frac{dc}{dz^*} = \frac{\dfrac{1-\varepsilon}{\varepsilon} \dfrac{\Delta q}{\Delta c} v}{1 + \dfrac{(1-\varepsilon)}{\varepsilon} \dfrac{\Delta q}{\Delta c}} \frac{dc}{dz^*} = U \frac{(1-\varepsilon)}{\varepsilon} \frac{\Delta q}{\Delta c} \frac{dc}{dz^*}$$

and:

$$\frac{(1-\varepsilon)}{\varepsilon} \frac{\partial q}{\partial t}\bigg|_z = -\frac{(1-\varepsilon)U}{\varepsilon} \frac{dq}{\partial z^*}$$

Finally:

$$D \frac{\partial^2 c}{\partial z^{*2}} = U \frac{(1-\varepsilon)}{\varepsilon} \left[\frac{\Delta q}{\Delta c} \frac{dc}{dz^*} - \frac{dq}{dz^*} \right] = \frac{U(1-\varepsilon)}{\varepsilon} \frac{\Delta q}{\Delta c} \left[\frac{dc}{dz^*} - \frac{\Delta c}{\Delta q} \frac{dq}{dz^*} \right]$$

We shall set:

$$x = \frac{c}{\Delta c} \quad \text{and} \quad y = \frac{q}{\Delta q} \quad \text{and divide by } \Delta c$$

$$D_a \frac{d^2x}{dz^{*2}} = \frac{U(1-\varepsilon)}{\varepsilon} \frac{\Delta q}{\Delta c} \frac{d(x-y)}{dz^*}$$

Let us integrate once:

$$\frac{dx}{dz^*} = \beta(x-y) \text{ where } \beta = \frac{U(1-\varepsilon)\Delta q}{\varepsilon D_a \Delta c} \qquad [2.6]$$

1) Finite resistance for the material transfer and $D_a = 0$

The transfer is written:

$$\left. \frac{\partial x}{\partial t} \right|_z = \left. \frac{\partial y}{\partial t} \right|_z = G(x,y) = k_y a(x^*(y)-y) = k_x a(x-y^*(x))$$

In addition:

$$\frac{dq}{dz^*} = \left. \frac{\partial y}{\partial t} \right|_z \times \left. \frac{\partial t}{\partial z^*} \right|_z = -\frac{G}{U} \qquad [2.7]$$

This equation can immediately be integrated.

2) Step of transfer resistance and $D_a > 0$

Equation (5) is written:

$$z^* = \frac{1}{\beta} \int_{x_0}^{x} \frac{dx}{x - y^*(x)}$$

3) Finite transfer resistance and $D_a > 0$

Let us divide equations [2.6] and [2.7], term by term:

$$\frac{dy}{dx} = \frac{-G}{\beta U(x-y)} = -\frac{k_y a(x^*(y)-y)}{\beta U(x-y)}$$

The integration of this equation gives y(x), which can be inserted into one of the two equations [2.6] or [2.7], which give $x(z^*)$ and $y(z^*)$.

4) [COO 65] offer detailed discussions about the existence of the solution in each case. [MIC 52] deals with the problem of the finite transfer resistance with $D_a = 0$ on the example of an exchanger resin $Na^+ \Leftrightarrow H^+$. [HAS 76, HAS 77] studied the case of a Freundlich adsorption isotherm. Hall *et al.* [HAL 66] examined the influence of intragranular diffusion on the constant profile. [LAP 54] looked at the problem of exchange of monovalent ions.

Finally, Rhee *et al.* [RHE 71] deal with the case where the isotherm contains an inflection point, as does the BET isotherm.

2.6.6. *Villermaux and Van Swaaij's [VIL 69] bell curve for a brief injection*

We suppose, as usual, that the contents in terms of solutes are measured in:

– concentration in the fluid phase c;

– also in concentration in the solid phase

$$q = \frac{\text{quantity of solute}}{\text{true volume of solid}}$$

The balance equation is:

$$u\frac{\partial c}{\partial z} + \varepsilon\frac{\partial c}{\partial t} + (1-\varepsilon)\frac{\partial q}{\partial t} = D_a \frac{\partial^2 c}{\partial z^2}$$

Let us multiply everything by Z/u, where Z is the height of the column and u is the velocity of the fluid in an empty bed:

$$Z\frac{\partial c}{\partial z} + \varepsilon\frac{Z}{u}\frac{\partial c}{\partial t} + (1-\varepsilon)\frac{\partial q}{\partial t}\frac{Z}{u} = \frac{D}{Zu}\frac{\partial^2 c}{\partial z^2}Z^2$$

We set:

$$x = \frac{z}{Z}; \quad \theta = \frac{ut}{Z}; \quad P = \frac{Zu}{D}$$

The balance equation becomes:

$$\frac{\partial c}{\partial x} + \varepsilon\frac{\partial c}{\partial \theta} + (1-\varepsilon)\frac{\partial q}{\partial \theta} = \frac{1}{P}\frac{\partial^2 c}{\partial z^2}$$

This equation was solved by [VIL 69], in the case where the feed is a Dirac distribution. These authors give the expression of the distribution function of the residence times at the output from the column. Let E (t) represent that distribution.

Between time t and time t + dt, the number dn of molecules exiting the column is:

$$dn = (Audt)\ c\ (t) = E\ (t)\ dt$$

Therefore:

$$E\ (t) = Auc(t)$$

Hence the result given by those authors:

$$c(t) = \frac{1}{Au}\sqrt{\frac{P}{\pi\theta}}\exp\left[-\frac{P(1-\theta)^2}{4\theta}\right]$$

$$-\frac{P}{2Au}(\exp P)\,\mathrm{erf}\left[\left(\frac{P}{\theta}\right)^{1/2}\frac{(1+\theta)}{2}\right]$$

With a sufficiently large value of θ, the argument of the error function tends toward $\frac{1}{2}(P\theta)^{1/2}$, meaning that erf tends toward 1.

Finally:

$$c(t) = \frac{1}{Au}\sqrt{\frac{P}{\pi\theta}}\exp\left[-\frac{P(-1-\theta)^2}{4\theta}\right] - \frac{P\exp P}{2Au}$$

We can see that the solution to the equation is not exactly given by a Gaussian curve. Instead, it is a bell curve whose standard deviation is given by equation 14 in [VIL 69].

$$\sigma^2 = \frac{2}{P} + \frac{3}{P^2}$$

Note, however, that these authors write:

$$\frac{\partial q}{\partial \theta} = \frac{kaZ}{\varepsilon uA}(c - q)$$

or indeed:

$$\frac{\partial q}{\partial t} = \frac{ka}{\varepsilon A}(c - q)$$

Here, there is an error, because there is no reason, with the hypothesis of a linear isotherm being adopted, for us to have:

c = aq where a = 1

The crux of the result remains valid, though.

2.7. Channeling of a single solute

2.7.1. Injection of any form. Abscissa of the discontinuity [DE 43]

We shall base our discussions on the work of [DE 43]. The injection is of the form:

c = g (z) or indeed z = S (c)

The material balance between the origin and the discontinuity is:

$$z_d \left[\varepsilon(c_b - c_a) + (1 - \varepsilon)(q_b - q_a) \right] = \int_{c_a}^{c_b} z \left(\varepsilon dc + (1 - \varepsilon) dq \right)$$

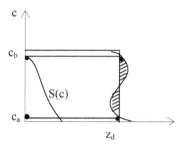

Figure 2.6. *Deformation of the concentration profile*

Yet:

$$z = S(c) + tU(c) = S(c) + \frac{ut}{1 + Ff'(c)}$$

and:

$$\varepsilon dc + (1-\varepsilon) dq = \varepsilon(1 + F f') dc$$

Finally:

$$z_d = \frac{ut(c_b - c_a) + \int_{c_a}^{c_b} (1 + Ff') S(c) dc}{c_b - c_a + F(f(c_b) - f(c_a))}$$

c_b and c_a are functions of z_d. Thus, this equation is implicit in z_d. However, with a rectangular pulse, we immediately obtain:

$$z_d = \frac{utc_0}{c_0 + Ff(c_0)} \qquad (c_a = 0 \text{ and } c_b = c_0)$$

2.7.2. Propagation of a rectangular pulse. Quasi-ideal system

In an ideal system where the isotherm is linear, all the concentrations propagate at the same speed, so the profile of the pulse moves along the column *without deformation*.

In a quasi-ideal system where the isotherm is Langmuirian, for strong concentrations, the value of f' is less than it is for weak concentrations.

$f'_{strong} < f'_{weak}$ which means that $U_{strong} > U_{weak}$

Figure 2.7(A) shows the deformation of the rectangular pulse in a Langmuirian isotherm.

Conversely, Figure 2.7(B) illustrates the deformation of the pulse for an antilangmuirian isotherm.

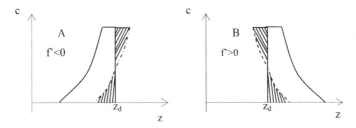

Figure 2.7. *Deformation of a rectangular pulse*

In Figure 2.7, the vertical line shows the position z_d of the center of gravity (barycenter) of the front of A and the back of B. However, on this vertical, there are three possible values for the concentration c, which is impossible, especially given that the solid too would have to be at equilibrium with each of the three concentrations. The only way to solve the problem is to hypothesize that there is discontinuity of the concentration at the abscissa z_d.

We can see that the amplitude of the discontinuity is $\Delta c = c_0 - 0$ for the fluid and $\Delta q = q_0 - 0$ for the solid.

More mathematically, we know that:

$$U = \frac{u}{1 + Ff'} \quad so \quad \frac{dU}{dc} = \frac{-u\,F\,f''}{\left(1 + F\,f'\right)^2}$$

Thus $f'' < 0$ $dU/dc > 0$ discontinuity in front

        $f'' > 0$ $dU/dc < 0$ discontinuity at back

2.7.3. *Propagation of a discontinuity in concentration [DE 43]*

Suppose that the discontinuity has moved by the length Δz corresponding to the volume of fluid $\varepsilon \Delta z$ and to the volume of solid $(1-\varepsilon) \Delta z$. The solute gained by the segment Δz is:

$$\left(c_{ia} - c_{ib}\right)\varepsilon\Delta z + \left(q_{ia} - q_{ib}\right)\left(1-\varepsilon\right)\Delta z$$

a and b: indices characterizing the medium upstream and downstream of the discontinuity.

The solute gained by the segment comes from the impoverishment of the fluid which, over the same time period, has moved by $\varepsilon u \Delta t$:

$$\varepsilon u \Delta t \left(c_{ia} - c_{ib}\right)$$

u: interstitial velocity of the fluid: $m.s^{-1}$.

By making these two quantities equal to one another, we obtain:

$$\frac{\Delta t}{\Delta z} = \left(1 + \frac{\left(1-\varepsilon\right)}{\varepsilon}\frac{q_{ia} - q_{ib}}{c_{ia} - c_{ib}}\right)\frac{1}{u}$$

Finally, the rate of displacement of the discontinuity is:

$$U = \frac{\Delta z}{\Delta t} = \frac{u}{1 + \dfrac{1-\varepsilon}{\varepsilon}\dfrac{\Delta q}{\Delta c}}$$

We can accept that, for all solutes, the discontinuities appear at the same place and at the same time. The result of this is that, for a discontinuity:

$$U_i = U_j \quad \text{and therefore that} \quad \frac{\Delta q_i}{\Delta c_i} = \frac{q_{ia} - q_{ib}}{c_{ia} - c_{ib}} = \frac{q_{ja} - q_{jb}}{c_{ja} - c_{jb}}$$

whatever the values of i and j, when:

$$1 \leq i \leq n \quad \text{and} \quad 1 \leq j \leq n$$

There is, however, one exception. If a solute has already been exhausted, obviously, it no longer plays a part in these equations.

2.7.4. Shape of the profile of q as a function of time (single solute)

Consider the case of a rectangular pulse, and let us see how the band of solute moves along the column as a function of time. We suppose the isotherm is favorable, meaning that $f''<0$.

A: profile of rectangular injection;

B: the front of the band is vertical but, on the other hand, the back of the band is progressive;

C: the front and back have come together, and the profile presents a peak;

D: the back meets the front earlier and the concentration q of the peak has decreased.

At the back of the band front, we use the term "train" (like a bridal train) to speak of the curve showing the progressive decrease in concentration.

Furthermore, if we bring additional phenomena into play, such as the axial dispersion, a non-instantaneous material transfer, and rounding errors in the numerical calculation, then the band takes the form of a bell curve.

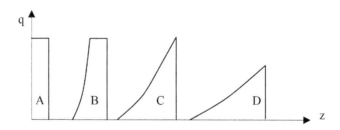

Figure 2.8. *Concentration profiles in the solid at successive times*

NOTE.–

The material balance equation is written:

$$\left(1+Ff'\right)\frac{\partial c}{\partial t}=-u\frac{\partial c}{\partial z}\qquad (\,f'\ \text{always positive})$$

For the same variation in concentration c_i, the variations of t and z are of the opposite signs. This gives us the shape of the profiles.

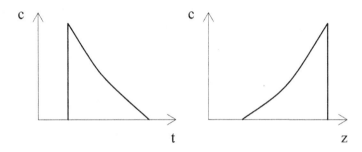

Figure 2.9. *Concentration profiles as a function of t and z quasi-ideal system (rectangular injection)*

These two profiles *cannot be deduced from one another* as mirror images.

It is important not to confuse the profile (over time) of elution at the output from the column and the profile (in space) inside of the column.

2.7.5. *Calculation of the elution time of the concentration front [GOL 88]*

After a rectangular injection whose duration is t_p, the front has traveled the distance z_f at the end of time t:

$$z_f = \frac{ut}{1+\dfrac{Fa}{1+bc_0}}\qquad\qquad [2.8]$$

The back only began to move at time t_p, and has traveled the distance:

$$z_a = \frac{u(t - t_p)}{1 + \dfrac{Fa}{(1 + bc_0)}} \qquad [2.9]$$

Expressions [2.8] and [2.9] show that the back progresses more quickly than the front. Indeed, in Figure [2.10], we can see that the slope of the tangent is less than that of the chord.

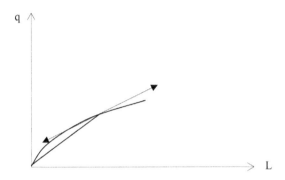

Figure 2.10. *Langmuirian isotherm*

The flat summit of the pulse therefore decreases in length, and disappears at the point z_i. From that moment on, the height of the discontinuity decreases and the concentration of the peak becomes a decreasing function of time.

The back catches up with the front when $z_a = z_f$. We can deduce the corresponding time t_I.

$$t_I = \frac{t_p(1 + bc_0)(1 + bc_0 + Fa)}{bc_0 Fa}$$

The common abscissa corresponding to that time is:

$$z_I = \frac{t_p u (1 + bc_0)^2}{Fa\, bc_0}$$

At the peak of the concentration profile, the variables t and c satisfy equations [2.10] and [2.11] respectively for the front and for the progressive variation.

Equation [2.9] can be written:

$$\frac{d\left(t - t_p - z/u\right)}{d\left(z/u\right)} = \frac{Fa}{1 + bc}$$

Equation [2.9] can also be written:

$$Fa\frac{\left(t - t_p - z/u\right)}{z/u} = \frac{Fa}{1 + bc}$$

Hence:

$$\frac{d\left(t - t_p - z/u\right)}{\left[Fa\left(t - t_p - z/u\right)\right]^{1/2}} = \frac{d\left(z/u\right)}{\left(z/u\right)^{1/2}}$$

$$t - t_p - z/u = \left[\left(Fa\, z/u\right)^{1/2} - const.^{1/2}\right]^2 \qquad [2.10]$$

Let us introduce the load ratio of the column, which is:

$$K = \frac{injected\ component}{capacity\ of\ the\ column\ at\ saturation\ for\ that\ component}$$

This means that:

$$K = \frac{c_0 \varepsilon A u t_p}{q_\infty \left(1 - \varepsilon\right) AH} \qquad [2.11]$$

c_0: concentration of the rectangular injection: kmol.m^{-3} or kg.m^{-3};

t_p: duration of the rectangular injection: s;

A: section of the column;

ε: empty fraction of the filling;

Z: height of the column;

q_∞: concentration of the solute in the solid: $kmol.m^{-3}$ or $kg.m^{-3}$;

u: interstitial velocity of the fluid: $m.s^{-1}$.

If the isotherm is Langmuirian:

$$q = \frac{ac}{1+bc} \qquad q_\infty = \frac{a}{b} \qquad \frac{dq}{dc} = \frac{a}{(1+bc)^2}$$

The length of stay of the fluid phase is:

$$t_{v0} = \frac{Z}{u}$$

The charge rate becomes:

$$K = \frac{c_0 t_p u b}{FaZ} = \frac{c_0 t_p b}{Fat_{v0}} \quad \text{where } t_{v0} = \frac{Z}{u}$$

By feeding t_l and z_l into equation [2.10], we obtain:

$$t_f = t_p + t_{v0} + Fa\, t_{v0}\left(1 - K^{1/2}\right)^2 \tag{2.12}$$

This equation gives the time t_f at which the discontinuity front passes the point of abscissa H. This is the elution time of the front.

The length of stay of a given concentration c on the progressive part of the profile is:

$$t = t_p + t_v\left(1 + \frac{Fa}{(1+bc)^2}\right)$$

Hence:

$$c = \frac{1}{b}\left[\left(\frac{t_{s0} - t_v}{t - t_p - t_v}\right)^{1/2} - 1\right]$$

From this, [GOL 88] deduce the concentration profile shown in Figure 2.11.

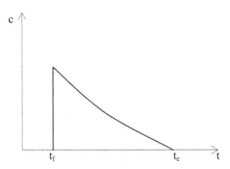

Figure 2.11. *Concentration profile at output for a quasi-ideal system*

In particular, they determine:

– the position of the maximum;

– the breadth of the profile at its base;

– the time t_f of passage of the front;

– the time t_e of passage of the end of the progressive part;

– the mid-height concentration on the profile.

2.7.6. *Position and breadth of the elution peak*

The moment μ_1' defines the mean moment of elution of the peak $c = f(t)$

$$\mu_1' = \frac{\displaystyle\int_0^\infty tc(z,t)\,dt}{\displaystyle\int_0^\infty c(z,t)\,dt}$$

The moment of order $2\mu_2$ defines the breadth of the peak of $c = f(t)$

$$\mu_2 = \frac{\displaystyle\int_0^\infty \left(t - \mu_1'\right)^2 c(z,t)\,dt}{\displaystyle\int_0^\infty c(z_i,t)\,dt}$$

[KUB 65a, KUB 65b] calculated the Laplace transforms of the concentrations in the equations of the system, and deduced the expression of the moments μ_1' and μ_2.

[SCH 68] reformulated Kubin's results in light of the following relations:

– material balance in the fluid phase;

– material balance in a particle;

– transfer at the surface of the particles;

$$D_s \left.\frac{\partial c_i}{\partial r}\right|_{r=R} = k_s \left(q_i^* - q_i\right)$$

q_i: concentration in the particle: $kmol.m^{-3}$;

q_i^*: concentration in the particle at equilibrium with the fluid: $kmol.m^{-3}$;

R: radius of the particle: m;

D_s: diffusivity of the solute in the particle

– rate of adsorption from the fluid toward the solid.

$$\frac{\partial c_{ads}}{\partial t} = k_s \left(q_i - \frac{q_{ads}}{K_A}\right)$$

K_A : equilibrium constant: dimensionless;

q_{ads}: mean concentration in the solid phase at the point (z, t): $kmol.m^{-3}$;

k_s: transfer coefficient: $m.s^{-1}$.

[SCH 68] formulation is as follows. They first posit:

$$\delta_0 = \frac{(1-\alpha)\beta}{\alpha}\left(1 + \frac{\rho_s K_A}{\beta}\right)$$

$$\delta_1 = \frac{(1-\alpha)\beta}{\alpha}\left[\frac{\rho_p K_A^2}{\beta k_{ads}} + \frac{R^2\beta}{15}\left(1 + \frac{\rho_p K_A}{\beta}\right)^2\left(\frac{1}{D_s} + \frac{5}{k_f R}\right)\right]$$

They then deduce:

$$\mu_1' = \frac{z}{u}(1+\delta_0) + \frac{t_0}{2}$$

$$\mu_2 = \frac{2}{u}\left[\delta_1 + \frac{D_a}{\alpha u^2}(1+\delta_0)^2\right] + \frac{t_0^2}{12}$$

α: intergranular porosity (void fraction in terms of volume);

β: internal porosity of each grain;

ρ_p: apparent density of a particle: $kg.m^{-3}$;

t_0: time needed for the saturation of the adsorbent for a transfer resistance of zero: s;

D_a: axial dispersivity: $m^2.s^{-1}$

$$t_0 = \frac{1-\alpha}{\alpha}\rho_p K_A \frac{z}{u}$$

u: interparticular velocity of the fluid: $m.s^{-1}$

$$u = U / \alpha$$

U: velocity of the fluid in an empty bed: $m.s^{-1}$.

2.8. Number of solutes greater than 1

2.8.1. *Study of a binary mixture [GOL 89]*

The above-named authors quantitatively studied elution curves at the output from an adsorption column. To do so, they used the Langmuirian isotherm adapted to two solutes (see Figure 2.12). They performed the operation in two separate columns:

– one very short column, at the output from which the two solutes are still partially mixed (Figure 2.12(a));

– one long column which assures the presence of two separate peaks for the two solutes (Figure 2.12(b)) at its output.

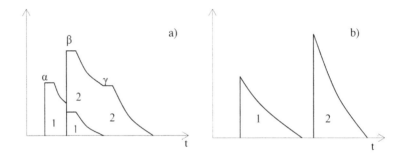

Figure 2.12. *Elution of a binary solution*

In Figure 2.12(a), we see the presence of three discontinuities.

– the discontinuity α separates the pure solvent from a solution containing only solute 1, which is least encumbered by the solid and therefore exits first;

– the discontinuity β is where solute 2 appears, solute 2 being the slower component in a solution of the two solutes. Thus, this discontinuity separates a solution of 1 and a solution of 1+2;

– the discontinuity γ corresponds to the disappearance of solute 1, which is to say that it separates a solution of 1+2 and a solution of 2.

In Figure 2.12(a), the discontinuities are followed by platforms at constant concentrations. However, the length of these platforms decreases and ultimately disappears as the fluid progresses through the column (when a taller column is used).

In Figure 2.12(b), the platforms have disappeared and the peaks corresponding to the two solutes have separated. This demonstrates the principle of chromatography.

Note that as soon as one of the solutes begins a platform, the other will immediately do likewise. The same is true when a platform completes.

2.8.2. Concentration profiles (multiple solutes) [TON 67, TON 68]

Consider a column of adsorbent subjected to the percolation of a fluid. Suppose that solutes are already present in the solid in the column, and suppose also that the fluid contains solutes (generally) different from the solutes of the solid.

The problem we now set ourselves is to find out how the concentrations in the solid appear the length of the column. We shall suppose that, at all points, the fluid is at equilibrium with the solid.

The total number of solutes is n, which varies between 2 and 4 and, more rarely, 5. We classify these solutes in order of decreasing affinity for the solid and represent them by a letter. Thus, in order of decreasing affinity, we would have:

A, B, C, D, E

Computer-based simulations have yielded the following results:

1) if a solute is present in the solid before the operation and is also present in feed fluid, then its concentration does not reach zero in the column;

2) if solutes are not present in one of the two phases, then their concentration falls to zero somewhere within the column. In the direction of progression of the fluid, the zero-concentration points are aligned in order:

A, B, C, D, E

3) the concentration profile appears as a series of platforms separated by transition zones, which may be progressive or may be sudden. If one

concentration changes suddenly, then the others do likewise. The number of platforms (including the initial concentrations of the fluid and solid) is equal to the number n of solutes. Thus, in the column, we find:

n platforms

n-1 transition zones

If all the solutes must disappear, then there comes a point of transition where two solutes reach zero concentrations.

The result is that, *a priori*, the concentration of the solute of index k can only fall to zero in one of the two transitions k or k-1. In other words, in the kth transition, only the solutes k or k+1 can disappear;

4) slope rule. The slopes in question are those of the concentrations between the platforms. A slope is said to be positive if the concentration of the solute at hand increases in the direction of progression of the fluid. Computer-based simulations have shown that, in order for a transition to be progressive, the slope of the solute in question must obey the following table:

	1	2	3	4
A	+	+	+	+
B	-	+	+	+
C	-	-	+	+
D	-	-	-	+
E	-	-	-	-

The digits are number of the transitions.

In a transition where a solute is depleted to nothing:

– if the slope has the sign corresponding to the table, the transition is progressive;

– if the slope has the opposite sign, the transition is sudden.

2.8.3. *Exceptions to the rule*

Consider a column of adsorbent subjected to the percolation of a fluid.

[TON 70] gave the conditions for which what he says an "inversion" may occur – put differently, when the concentrations do not reach zero in the order of the affinities of the solutes for the solid.

2.9. Numerical simulation by orthogonal collocation. Finite-element method

2.9.1. *Presentation of the problem*

We shall base our discussion on the publications of, firstly, [VIL 67] and, secondly, [YU 89]. Readers could usefully consult these two documents. The goal is to find an expansion into a series of even-numbered polynomials (in x^2) for a function y which must satisfy a second-order linear differential equation (in y).

$$y = \alpha_0 + \sum_{i=1}^{n} \alpha_i p\left(x^2\right) P_i\left(x^2\right)$$
$$y'' A(x) + y' B(x) + y\, C(x) + D(x) = 0$$

$p(x^2)$ is a weight function whose form depends on the type of polynomial chosen.

For such a differential equation, there are conventional numerical solution methods, e.g. the finite-difference method, which we shall see later on.

However, the absorption of a liquid or gaseous component on a bed of solid particles results in sudden variations of the profile of concentration of the solid along the column and of the concentration of the output fluid as a function of time.

This *rigidity* of the profiles necessitates the use of orthogonal polynomials, because this is the only numerical method which always delivers perfect precision (five or six meaningful digits) for the concentrations. In addition, the orthogonality of the polynomials makes for a simple method to calculate the coefficients.

NOTE.–

When a variable z traverses the interval (A, B), we introduce the variable u which, for its part, varies within the interval (a,b).

$$u = \frac{\left[(B-A)z + Ab - aB\right]}{b-a}$$

In general:

a = 0 and b = 1 or else a = 1 and b = 0

This enables us to use the normalized forms of the polynomials.

2.9.2. *The finite-element method*

The column is divided into four to eight segments (the fixed elements), all of which have the same thickness and within each of them, we perform an orthogonal-polynomial collocation. The concentrations and their first spatial derivatives are continuous from one element to an other.

[YU 89] propose a method to take account of the rigidity of concentration profiles. They split each element into sub-elements, the number of which increases with the slope of the concentration curves.

As regards the evolution of the concentration c_s within the solid particles, we use an expansion in two or three polynomials and a single element along the length of the radius of the particle.

2.9.3. *Orthogonal polynomials*

If we seek the solution to the above differential equations in the form of a single polynomial whose coefficients of the powers of x are to be determined, experience proves that the accuracy of the polynomial function obtained is most mediocre.

On the other hand, the problem has a very precise solution if we look for the coefficients a_i such that:

$$y(x) = \alpha_0 + \sum_{j=1}^{n-1} \alpha_j p(x) P_j(x)$$

The P_j are the polynomials.

$$P_j(x) = \alpha_0 + \alpha_1 x + \alpha_2 x^2 + a_j x_j + \alpha_n x^n$$

So that we can determine the coefficients α_j easily, the polynomials must be orthogonal to one another, meaning that:

$$\int_0^1 p(x) P_j(x) x^{a-1} dx = C_j \delta_{jk}$$

δ_{jk} : Kronecker delta;

p (x): weight function (equal to $1-x^2$ for the Jacobi polynomials).

There are many different families of orthogonal polynomials, and for each of these families, there is a corresponding particular weight function. These weight functions are as follows, depending on the nature of the polynomials.

Polynomials	Weight function
Jacobi	$1 - x^2$
Legendre	1
Chebyshev	$\left(1 - x^2\right)^{-1/2}$

Table 2.3. *Nature of weight functions*

The necessary details about polynomials are to be found in [SPI 74] and, more specifically, details about Jacobi polynomials are given in [VIL 67].

Thus, we can deduce the coefficients α_j.

$$\int_0^1 y(x)p(x)P_j(x)x^{a-1}dx = \alpha_j \int_0^1 p(x)P_j(x)P_j(x)x^{a-1}dx = C_j\alpha_j$$

$$\alpha_j = \frac{1}{C_j}\int_0^1 y(x)p(x)P_j(x)x^{a-1}dx$$

2.9.4. *Error in serial expansion: collocation nodes*

For practical reasons, we cannot indefinitely continue a serial expansion. Suppose that the last term taken into account is the (n-1)-order term.

The difference between the series truncated at order (n-1) and the infinite series is the truncation error ε.

$$\varepsilon = \sum_{j=n}^{\infty} p(x)P_j(x)$$

Naturally, we suppose the series to be convergent, which means that the modulus of the terms decreases with their order.

The most important term in this sum is therefore the n-order term, and a way to minimize ε is therefore to make that term equal to 0, which occurs for values of x equal to the roots of the polynomial p(x) $P_n(x)$.

Those roots are known as the nodes of the collocation support (from the Latin *collocare*: to put in place). At these points, the accuracy of the serial expansion is maximal, so it is there that we shall study the evolution of the concentrations over time.

2.9.5. *Discretization matrix [A]*

In the interval (0,1), the variable x takes n values:

with $x_{n+1} = 1$

$$0 \quad x_1 \quad\quad\quad x_{i-1} \quad x_n \quad\quad 1$$

We shall evaluate the matrix [A]:

$$[A] = \begin{pmatrix} A_{11} & A_{1i} & A_{1n+1} \\ A_{k1} & A_{ki} & A_{k,n+1} \\ A_{n+1,1} & A_{n+1,i} & A_{n+1,n+1} \end{pmatrix}$$

Let us now construct the matrix of even-numbered powers of x for x varying from x_1 to x_{n+1}.

$$\left[X^2 \right] = \begin{pmatrix} 1 & x_1^{2j} & x_1^{2n} \\ 1 & x_i^{2j} & x_i^{2n} \\ 1 & x_{n+1}^{2j} & x_{n+1}^{2n} \end{pmatrix}$$

The matrix of the first derivatives of x^{2j} is:

$$\left[DX^2 \right] = \begin{pmatrix} 0 & \dfrac{dx^{2j}}{dx}\bigg|_{x_1} & \dfrac{dx^{2n}}{dx}\bigg|_{x_1} \\ 0 & \dfrac{dx^{2j}}{dx}\bigg|_{x_i} & \dfrac{dx^{2n}}{dx}\bigg|_{x_i} \\ 0 & \dfrac{dx^{2j}}{dx}\bigg|_{x_{n+1}} & \dfrac{dx^{2n}}{dx}\bigg|_{x_{n+1}} \end{pmatrix}$$

The determinant of these matrices is supposed to be different to zero.

The matrix [A] is determined by the relation:

$$\left[DX^2 \right] = [A] \cdot \left[X^2 \right] \qquad\qquad [2.13]$$

If $[X^2]^{-1}$ is the inverse of the matrix $[X^2]$, we have:

$$\left[X^2 \right] . \left[X^2 \right]^{-1} = [I] \text{ and } [A].[I] = [A] \left([I] \text{ is the identity matrix} \right)$$

Hence, by multiplying both sides of equation [2.13] by $[X^2]^{-1}$:

$$[A] = \left[DX^2 \right] . \left[X^2 \right]^{-1}$$

2.9.6. *Generalization to other discretization matrices*

By a similar approach, we can extend the above result to other linear operators:

Second derivative:

$$\frac{d^2 x^{2j}}{dx}\bigg|_{x_i}$$

and definite integral:

$$\int_0^1 x^{2j+a-1} dx = s_j$$

Hence:

$$[B] = [D^2 X^2] \cdot [X^2]^{-1}$$

[2.14]

$$[W] = (S) \cdot [X^2]^{-1}$$

(S) is a row vector:

$$\alpha_k = \frac{1}{C_k} \sum_{j-1}^{n} W_j [y(x_i) - \alpha_0] P_k (x_j^2)$$

The values of the C_k are to be found in [VIL 67]. These values are given by their equation 9 for the Jacobi polynomials.

2.9.7. *Case of any given even-numbered polynomial*

Such a polynomial is of the form:

$$y = \sum_{j=0}^{n} a_j x^{2j}$$

This expression can be considered to be the product of two vectors:

– the row vector $(1, x^2, ..., x^{2j}, ..., x^{2n}) = (X^2)$;

– the column vector

$$(a) = \begin{pmatrix} a_0 \\ a_j \\ a_n \end{pmatrix}$$

The result is a scalar (it is the scalar product of the two vectors).

Thus:

$$y = (X)^2 . (a)$$

If we successively give the values x_i to the variable x, the function y becomes a vertical vector (y):

$$(y) = \left[X^2 \right] . (a)$$

Similarly:

$$(y') = \left[DX^2 \right] . (a)$$

$$(y'') = \left[D^2 X^2 \right] . (a)$$

Equations [2.13] and [2.14] become the following when we multiply both sides by the column vector (a):

$$(y') = [A](y)$$

$$(y'') = [B](y)$$

The row with index i is written:

$$\left. \frac{dy}{dx} \right|_{x_i} = \sum_{k-1}^{n-1} A_{ik} \, y \big|_{x_k}$$

EXAMPLE 2.4.–

Let us choose the Jacobi polynomial P_1 (x^2) for which $n = 1$

$$P_1\left(x^2\right) = 1 - 5x^2$$

The positive root x_1 is $\dfrac{1}{\sqrt{5}}$

The other positive root is $x_2 = 1$ because:

$$y = y\big|_{x=x_1} + \left(1 - x^2\right)a_1 P_2\left(x\right)$$

We can verify that, by replacing x_1 and x_2 with their values, if we have:

$$\left[X^2\right] = \begin{bmatrix} 1 & x_1^2 \\ 1 & x_2^2 \end{bmatrix} \quad [A] = \begin{bmatrix} \dfrac{-5}{2}x_1 & \dfrac{5}{2}x_1 \\ \dfrac{-5}{2} & \dfrac{5}{2} \end{bmatrix} \quad \text{(see [VIL 67])}$$

we do indeed have:

$$\left[\dfrac{\partial x^{2j}}{\partial x}\right] = [A].\left[X^2\right] = \begin{vmatrix} 0 & 2x_1 \\ 0 & 2 \end{vmatrix} = \begin{vmatrix} \dfrac{\partial x_0}{\partial x}\bigg|_{X_1} & \dfrac{\partial x^2}{\partial x}\bigg|_{X_1} \\ \dfrac{\partial x_0}{\partial x}\bigg|_{X_2} & \dfrac{\partial x^2}{\partial x}\bigg|_{X_2} \end{vmatrix}$$

NOTE.–

[MIC 72] proffer a different method for calculating the discretization matrices.

2.9.8. *Integration with respect to time*

If, in the differential equations, we replace the first and second derivatives with their discretized forms, we obtain two systems of equations, which are both of the form:

$$\frac{du_i}{d\tau} = f\left(u_1, ..., u_i, ..., u_{n+1}\right)$$

The number n + 1 is not necessarily the same for:

$u_i = C_i$ and $u_i = C_{si}$

That number n + 1, remember, is the number of roots of the polynomial P_{n+1}.

The exchange equation solidarizes the two systems of equations which must therefore be integrated *simultaneously* as a function of the time. Indeed, we know that, in this equation, the following play a role:

$$c_j \quad \text{and} \quad c_j^* = f\left(c_{si}, ..., c_{sj}, ..., c_{sk}\right)$$

k: number of solutes.

This last equation is simply the equation of the isotherm.

Integrations as a function of time may be performed with the 4th-order Runge–Kutta method generalized to (n+1) unknowns (see Appendix 1).

NOTE.–

The equations of the k solutes must be integrated *simultaneously* so that the equation of the isotherm be coherent.

NOTE.–

The concentration and its evolution over time within the solid particles need to be calculated separately for each value x_i in each segment of the column.

2.10. Other numerical methods

2.10.1. *Finite-difference method*

The conditions are as follows:

– local equilibrium, the isotherm is not necessarily linear;

– presence of axial dispersion.

This is expressed by the balance equation:

$$u\frac{\partial c}{\partial z}+\left(1+F\frac{dq}{dc}\right)\frac{\partial c}{\partial t}=D_a\frac{\partial^2 c}{\partial z^2}$$

where:

$$F=\frac{1-\varepsilon}{\varepsilon}$$

Primarily, two possibilities arise, which we shall now examine.

1) From the input of the fluid into the column, determination of the concentration profile at time t on the basis of the profile at time t - Δt.

In the finite-difference method, the balance equation is written as follows, with axial dispersivity being ignored:

$$u\frac{\left(c_{z,t}-c_{z-1,t}\right)}{\Delta z}+\left(1+F\frac{dq}{dc}\right)\frac{\left(c_{z-1,t}-c_{z-1,t-1}\right)}{\Delta t}=0$$

Let us multiply all the terms by $\dfrac{\Delta t}{1+F\dfrac{dq}{dc}}$. We find:

$$\frac{u\Delta t}{\left(1+F\dfrac{dq}{dc}\right)}\left(c_{z,t}-c_{z-1,t}\right)+\left(c_{z-1,t}-c_{z-1,t-1}\right)=0$$

The Courant number P is defined by:

$$P = \frac{u\Delta t}{\left(1 + F\dfrac{dq}{dc}\right)\Delta z}$$

The first term in the balance equation is written:

$$c_{z,t} = \left(1 - \frac{1}{P}\right)c_{z-1,t} + \frac{1}{P}c_{z-1,t-1}$$

This formulation is valid only if:

$$P \geq 1 \text{ meaning that } \frac{u\Delta t}{\Delta z} \geq 1 + F\left(\frac{dq}{dc}\right)_{max}$$

(e.g. P = 2).

2) We calculate the concentration profile at time $t + \Delta t$ on the basis of the profile at time t.

In the finite-difference method, the balance equation is written as follows, with the right-hand side being ignored:

$$\left(c_{z,t} - c_{z,t-1}\right) + \frac{u\Delta t}{\left(1 + F\dfrac{dq}{dc}\right)\Delta z} + \left(c_{z,t-1} - c_{z-1,t}\right) = 0$$

or indeed:

$$c_{z,t} = \left(1 - P\right)c_{z,t-1} + P\,c_{z-1,t-1} = 0$$

This formulation is legitimate only if:

$$P \leq 1 \text{ meaning that } \frac{u\Delta t}{\Delta z} \leq 1 + F\left(\frac{\partial q}{dc}\right)_{min}$$

(e.g. P = 0.5).

If we take account of axial diffusion, for example, in case 1, the balance equation is written:

$$\frac{u\Delta t}{\left(1+F\dfrac{dq}{dc}\right)dz}\left(c_{z,t}-c_{z-1,t}\right)+\left(c_{z-1,t}-c_{z-1,t-1}\right)=\frac{D_a\Delta t}{\left(1+F\dfrac{dq}{dc}\right)\Delta z^2}\left(c_{z+1,t-1}-2c_{z,t}+c_{z-1,t-1}\right)$$

[CZO 90] put forward a relation which enables us to choose Δz when we know P (specified by [GUI 06]).

case a) $\Delta z(P\text{-}1)=\dfrac{2D_a}{u}$

case b) $\Delta z\,(1\text{-}P)=\dfrac{2D_a}{u}$

2.10.2. Craig's method

The sample is transferred into the fluid in the first cell. After equilibrium is reached between the two phases in that cell, the fraction of solute present in the fluid is r and is deduced from the total mass of solute in the two phases. We then express r as a function of k_0 for a polynomial.

The fluid phase then passes into the second cell, and at the same time, "fresh" fluid is added into the first one. The two cells are then balanced by calculating r_1 and r_2.

Then, fresh fluid is added into cell 1 after the fluid in 1 has flowed into 2 and the fluid in 2 has passed into 3.

The process is repeated until 99.999% of the injected sample has left the last cell (whose index is n).

Characterize each concentration by two indices and an exponent: thus, $c_{j,i}^k$ means the concentration of the solute i in the cell j at iteration k. The material balance is written:

$$(1-\varepsilon)q_{j,i}^{k-1}+\varepsilon\,c_{j-1,i}^{k-1}=\varepsilon\,c_{j,i}^{k}+(1-\varepsilon)q_{j,i}^{k}$$

Furthermore, the equilibrium of the cell j is written:

$$q_{j,i}^{k} / c_{j,i}^{k} = f\left(c_{j,1}^{k}, ..., c_{j,i}^{k}, ..., c_{j,n}\right)$$

Here, n is the number of solutes. Thus, for the iteration k and the cell j, we have n material balance equations and n equilibrium equations, which means we can obtain the n $q_{j,i}^{k}$ and the n $c_{j,i}^{k}$. We can see that, depending on the form of the function f, this calculation may not be simple.

2.11. Ion exchange

2.11.1. *Principle of ion exchange*

Ion exchange involves eliminating the solution of inconvenient ions by fixing them to a bed (generally fixed) of "exchanger" resins. However, the resin releases ions that do not present a disadvantage into the solution. This is the active phase.

When the resin reaches saturation, it is necessary to regenerate it with a so-called "regeneration solution": this is the regeneration phase. Ion exchange is also known as "permutation" (of ions).

2.11.2. *Exchange capacity – units used*

The unit of concentration (water hardness) is the French degree.

By definition, the French degree is abbreviated as °f, with a lowercase f so as not to confuse it with the degree Fahrenheit. The French degree corresponds to 0.2 milliequivalents per liter.

$$1°f = 0.2 \text{mequ.L}^{-1} = 0.2 \text{equ.m}^{-3}$$

Let us bring in the kiloequivalents per m³.

$$1 \text{ kequ.m}^{-3} = 5000°f$$

$$1°f = \frac{1}{5000} \text{kequ.m}^{-3}$$

Consider a volume V_L m^3 of a liquid solution containing ions to be permuted at the concentration q°f. The product V_Lq corresponds to the exchange of a number of kequ equal to V_Lq/5000. This operation requires the saturation of a volume V_R of exchanging solid material. The exchange capacity of that material is then:

$$\frac{V_L q}{5000 V_R} \text{ in equivalent kilomoles per apparent m}^3 \text{ of resin}$$

or indeed:

$$V_L q / V_R \text{ in °f.m}^{-3} \text{ apparent of resin}$$

For example:

Exchanging solid material	Exchange capacity	
	°f	kequ.m^{-3}
Natural clay	1000	0.2
Carboxylic resin	12000	2.4
Sulfonic resin	7000	1.4
Anionic resin (mildly basic)	5500	1.1
Anionic resin (strongly basic)	4000	0.8

Table 2.4. *Exchange capacities*

2.11.3. *Weakly-acidic or weakly-basic resins*

We can draw a distinction between:

a) Cationic resins, which exchange weakly-acidic cations containing carboxylic radicals grafted to their skeleton. Indeed, we know that carboxylic acids are weak acids.

These resins can be used in simple softening:

$$2RNa+Ca(HCO_3)_2 \xrightarrow{\text{Desalination}} R_2Ca+2NaHCO_3$$

$$R_2Ca+2NaCl \xleftarrow{\text{Regeneration}} 2RNa+CaCl_2$$

This softening can only pertain to weakly-bound cations, i.e. cations that are bonded to weak acidic anions – in other words, bicarbonates. We can therefore say that, on its own, the temporary hardness is eliminated.

Weakly-acidic resins can be regenerated using hydrochloric acid instead of NaCl salt. Thus, it is a decarbonation operation. Indeed, carboxylic resins fix the Ca^{++} and Mg^{++} ions, but also (to a lesser extent) the weakly-bound Na^+ ion, which is acidic sodium carbonate:

$$RH+NaHCO_3 \xrightarrow{\text{Decarbonation}} RNa+H_2O+CO_2$$

$$RNa+HCl \xrightarrow{\text{Regeneration}} RH+NaCl$$

During decarbonation, the free CO_2 produced may be released and leave the solution if the temperature and pH are favorable.

b) Anionic resins which are weakly-basic anion exchangers containing amine groups on their polystyrene skeleton. Indeed, amines are weak bases:

$$RNH_2+H_2O \rightleftarrows RNH_3^+ +OH^-$$

These resins exchange the anions of weakly-basic salts such as lime:

$$2RNH_3Cl+Ca(OH)_2 \xrightarrow{\text{Decalcification}} 2RNH_2+H_2O+CaCl_2$$

$$RNH_2+HCl \xrightarrow{\text{Regeneration}} RNH_3Cl$$

These resins are delivered in chloride form. They are insufficiently basic to fix anions of weak acids (silicic acid and carbonic acid).

However, decarbonation is more often done with carboxylic resins than with amine resins.

2.11.4. *Strongly-acidic or strongly-basic resins*

These resins are essentially polystyrene chains linked by divinylbenzene bridges. We can distinguish between:

a) Strongly-acidic cationic resins which exchange strong or weak cations, meaning cations of alkalis or alkali-earth metals. The sulfonyl functional groups grafted to the skeleton of the resin are strongly acidic, because their protons are mobile:

$$2RSO_3H+Ca(HCO_3)_2 \xrightarrow{\text{Decarbonation}} Ca(RSO_3)_2+2H_2O+2CO_2$$

$$(RSO_3)_2Ca+2HCl \xrightarrow{\text{Regeneration}} CaCl_2+2RSO_3H$$

Regeneration can also be carried out using the salt NaCl if we are looking for simple desalination.

b) Strongly-basic anionic resins, which fix strong or weak anions. The skeleton of the resin includes quaternary ammonium groups:

$$2NR_4OH+Na_2SO_4 \xrightarrow{\text{Desulfatation}} (NR_4)_2SO_4+2NaOH$$

$$(NR_4)_2SO_4+2NaOH \xrightarrow{\text{Regeneration}} 2NR_4OH+Na_2SO_4$$

If the regeneration is performed with NaCl, we obtain the resin in the chloride form.

Strongly-basic resins also fix weak anions such as the silicic anion SiO_3^{2-} or the carbonic anion CO_3^{2-}.

These resins are delivered in the chloride form or in the form of a free base. They are particularly vulnerable to organic debris, to which they fix irreversibly. Indeed, these materials carry a negative charge, because their surface contains amino acids which have lost their proton.

2.11.5. *Regeneration liquors*

Solutions of acids or bases, as we know, contain anions and cations which have different roles when it comes to ion permutation. Indeed:

– a proton corresponds to a strong acid if the acid in question is totally (or almost entirely) dissociated in an aqueous solution. A weak acid is only partially dissociated. Thus, a strong acid will provide the resin with a much higher number of protons and, therefore, the proportion of H^+ fixed by the resin will be very high. Cationic resins preferentially fix protons from strong acids;

– hydroxides, for their part, are always totally dissociated but are more or less soluble depending on the value of their solubility product. Consequently, caustic soda, which is highly soluble, presents far more OH^- ions than lime does, which is not very soluble.

This is why resins are regenerated using hydrochloric acid (a strong acid) and caustic soda (a strong base).

2.11.6. *Partition and selectivity*

The affinity of a resin for a cation increases with the valence of the ion. Figuratively, we could say that the resin can more easily catch the ion with two or more arms than with only one. Thus, alkali earth ions (Ca^{++} and Mg^{++} in particular) are more easily fixed than alkali ions (essentially Na^+ and K^+).

At equal valences, the affinity of the resins for an ion is stronger when the ion is less bulky, i.e. if its atomic mass is high. Indeed, the electrical attraction of nuclei with large atomic mass for the electron cloud is intense, and therefore the ionic radius is decreased. Thus, *in order of decreasing affinity*, ions are classed as follows:

a) monovalent cations:

Rb^+, NH_4^+ or K^+, Na^+, Li^+

b) bivalent cations:

$$Ba^{++}, \; Sr^{++}, \; Ca^{++}, \; Ni^{++}, \; Cu^{++}, \; Ca^{++}, \; Zn^{++}, \; Mg^{++}$$

(potassium is better fixed than sodium, and calcium is more easily captured than magnesium);

c) anions:

$$SO_4^{2-}, \text{ citrate, tartrate, } NO_3^-, \; PO_4^{3-}, \text{ acetate, } I^-, \; Br^-, \; C\ell^-, \; F^-$$

Anions of weak acids in aqueous solution are less easily fixed than anions of strong acids for the same reasons as discussed as regards regeneration liquors. Thus, the weak acid anions CO_3^{2-} and SiO^{2-} are difficult to fix. They are captured only by strongly basic resins.

The affinity of the resins for the H^+ ion is weak when the resin is strongly acidic.

The affinity of the resins for the OH^- ion is weaker when the resin is strongly basic.

The affinities can be evaluated using the partition coefficients.

The partition coefficients of a resin relative to a given ion i is:

$$m_i = \frac{q_i}{c_i} \rho_a$$

For the ion j:

$$m_j = \frac{q_j}{c_j} \rho_a$$

The selectivity of the resin is relative to the ion pair i, j and:

$$S_{ij} = \frac{m_i}{m_j} = \frac{q_i}{q_j} \times \frac{c_j}{c_i}$$

c_i: concentration of the compound in the liquid solution: $kg.m^{-3}$ or $kiloequivalent.m^{-3}$;

ρ_a: apparent density of the resin after swelling: $kg.m^{-3}$;

q_i: content of the resin in kiloequivalent per kg of resin ($kequiv.kg^{-1}$);

$\rho_a q_i$: concentration of the compound i in the apparent volume (after swelling) of the resin: $kequiv.m^{-3}$.

The classification in order of decreasing affinity corresponds to the staging of the different compounds taken in the direction of percolation, because the resin first fixes those products for which it has the greatest affinity.

Leakages

During the acid regeneration of a cationic resin, the solution which reaches the bottom of the bed of resin in contact with the sodium is already impoverished in terms of H^+, so the Na^+ ions are only incompletely extracted from the resin. When the resin is activated once more, the water acquires a growing level of acidity (as it percolates through to the bottom of the bed), because of the replacement of the Ca, Mg, Na and K cations by H^+. However, at the bottom, H^+ ions from the water will be replaced by Na^+ ions, and we see leakage of Na^+ ions with the solution, with formation of sodium silicates and carbonates if the water contained dissolved silica and CO_2.

If that water is then treated with a basic resin, the leakage of silica and CO_2 will be greater, because the neutral silicates and carbonates of sodium are more difficult to exchange than the corresponding acids.

In a mixed bed, we see the *simultaneous* depletion of the solution in both cations and anions, and the sodium leakage is non-existent.

2.11.7. *Exchange capacities*

We can distinguish the theoretical exchange capacity, linked to the chemical formula of the resin, and the practical capacity, which is lower, because total saturation of a bed of resin is never achieved by the time regeneration commences. This is due to the sigmoidal form of the penetration curve.

The practical exchange capacity does not grow proportionally to the mass of regeneration reagent used. We can simply state that the greater this mass is, the less the leakage will be.

Table 2.5 gives an approximate idea of the theoretical capacities of resins.

Type of resin	Theoretical exchange capacity (°f)		
strong cationic	8000	to	11000
weak cationic	13000	to	20000
strong anionic	5000	to	7000
weak anionic	5000	To	8000

Table 2.5. *Theoretical exchange capacities*

NOTE.–

The level of regeneration (acid, caustic soda or NaCl) is, in summary:

1.2 to 1.3 for weak resins,

2 to 3 for strong resins.

The product is first processed on a couple of primary columns filled with weak resins and then on a couple of finishing columns filled with strong resins and whose volume is slight because it is only a question of perfecting the exchange operation.

The regeneration liquor first passes over the strong resin and then over the weak resin, and thus we find the above levels of regeneration as long as the capacity of the weak and strong columns is in the ratio:

2/1.2 to 3/1.3

Weak resins (acidic or basic) may make do with a level of regeneration of 1.1 to 1.3 with respect to the stoichiometry of the ion mass fixed. Strong resins may also accept a moderate level of regeneration on condition that it increases the volume of the resin. Indeed, caustic soda and the acid HCl do not come for free. For softeners regenerated with NaCl with a regeneration rate of 3, we recover the second half of the regeneration solution (the last 50%) and reuse it in the next operation. Indeed, that recovered half remains rich in unused reagent. Certain authors put forward the following table.

Consumption (g of NaCl per °f.m³)	Practical exchange capacity (°f)
35	6600
30	6000
25	5000
18 (recovery of 50%)	5000
45 (excess speed of passage)	5000

Table 2.6. *Practical exchange capacities*

2.11.8. *Exchange of ions of the same valence [THO 44]*

$$A^{a+} + B \, \text{Resin} = B^{a+} + A \, \text{Resin}$$

$$\frac{dq}{dt} = k_1^* (Q - q) c - k_2^* (c_0 - c) q$$

Hereafter, we shall set:

$$k_1 = k_1^* F \qquad k_2 = k_2^* F \quad \text{where} \quad F = \frac{1 - \varepsilon}{\varepsilon}$$

The isotherm is given by the law of mass action:

$$\frac{k_2}{k_1} = \frac{(Q - q) c}{(c_0 - c) q} = K$$

$$Ay = \frac{k_2 c_0 (V - \varepsilon v)}{\dot{V}} \qquad\qquad Bx = \frac{k_1 Q \varepsilon v}{\dot{V}}$$

$$\alpha y = \frac{k_1 c_0 (V - \varepsilon v)}{\dot{V}} \qquad\qquad \beta x = \frac{k_2 Q \varepsilon v}{\dot{V}}$$

For example, in the case of adsorption:

$$\frac{c}{c_0} = \frac{I_0\left(2\sqrt{ABxy}\right) + \varphi(\alpha y, \beta x)}{I_0\left(2\sqrt{ABxy}\right) + \varphi(Bx, Ay) + \varphi(\alpha y, \beta x)}$$

$$\frac{q}{Q} = \frac{\varphi(\alpha y, Bx)}{I_0\left(2\sqrt{ABxy}\right) + \varphi(Bx, Ay) + \varphi(\alpha y, \beta x)}$$

V: volume fed in from time zero: m^3;

εv: volume available for the liquid in the length z of the column: n;

S: section of the column: m^2;

$\dot{V}$: flowrate of fluid fed in: $m^3.s^{-1}$.

2.11.9. Coefficient of the reaction kinetics [NEL 56]

The exchange of ions of the same valence is expressed by the relation:

$$\frac{dq}{dt} = \frac{\kappa}{Q}\left[c(Q-q) - \frac{1}{K}q(c_0 - c)\right]$$

κ: kinetic constant (pronounced "kappa"): s^{-1};

K: equilibrium constant: dimensionless.

The constant κ is given by:

$$\frac{ab}{\kappa} = \frac{1}{k_f} + \frac{c_0}{Qk_p}$$

The transfer coefficients k_f and k_p are given in Chapter 3 of [DUR 16c] and Chapter 4 of this book, and are measured in $m.s^{-1}$.

a: volumetric surface of the solid: $m^2.m^{-3}$

b: is a dimensionless constant that is a function of $r = 1/K$ and of the parameter ζ ("zeta").

$$\zeta = \frac{Qk_p}{c_0 k_f}$$

The authors give analytical expressions for b and, in particular, a family of curves $b = f(\zeta)$ parameterized in r.

To form the link with the calculation performed by Thomas [THO 44], we simply need to set:

$$k_1^* = \frac{\kappa}{Q} \quad \text{and} \quad k_2^* = \frac{\kappa}{QK}$$

NOTE.–

[HIE 52] give lattices of curves c/c_0 as a function of t/s with s as a parameter. Each lattice of curves corresponds to a value of the parameter r.

Type of sorption	r	T	S
ion exchange	k_2 / k_1	$\dfrac{k_1 c_0 (V - \varepsilon v)}{\dot{V}}$	$\dfrac{\varepsilon v}{\dot{V}}$
Langmuir	$\dfrac{k_2}{k_2 + k_1 c_0}$	$\dfrac{c_0 k_1 + k_2 (V - \varepsilon v)}{\dot{V}}$	$\dfrac{\varepsilon v}{\dot{V}}$

Table 2.7. Hiester and Vermeulen's parameter

[HIE 52] did not discuss the case of the linear isotherm.

[VER 52] precisely define the general meaning of parameters t and s.

NOTE.–

If the fluid flowrate in $\dot{V}$ tends toward zero, thermodynamic equilibrium is established between the solid and fluid. [GOL 83] studied the asymptotic behavior of the function used by [HIE 52].

$$J(x,y) = 1 - e^{-x-y}\varphi(x,y)$$

[GOU 65] studied the asymptotic behavior of the Bessel functions.

Practical Data on Adsorption, Ion Exchange and Chromatography

3.1. A few characteristics of fluid–solid equilibrium

3.1.1. *Fluid–solid equilibria*

Figure 3.1 shows the shape of the isotherms pertaining to water vapor and the most common adsorbents.

Let us now add a few practical specifications.

1) Ion-exchange resins

The capacity of a resin must be measured in $keq.m^{-3}$ (i.e. in eq. L^{-1}). It is important to specify whether the volume in question is reported before or after swelling.

In order of increasing affinity for the adsorbent, ions are classified as follows:

$$Li^+, Na^+, K^+, Rb^+, Ca^{++}, Mg^{++}, Ca^{++}, Sr^{++}, Y^{+++}, La^{+++}, \text{etc.}$$

F^-, Cl^-, Br^-, I^-, $CH_3 COO^-$, molybdate, phosphate, arseniate, NO_3^-, citrate, chromate, sulphate.

For OH^-:

– if the resin is weakly basic, OH^- is placed after SO_4^{--};

– if the resin is strongly basic, OH^- is placed before F^-.

2) Gases

The value of q increases if the pressure increases and if the temperature decreases, the odors are volatile, and their adsorption is more easily done at a lower temperature.

Desorption (regeneration of an adsorbent) can be performed:

– at reduced pressure;

– at high temperature.

Charcoal may, like silica, exhibit hysteresis with water vapor.

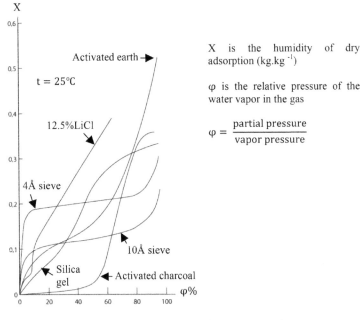

X   is   the   humidity   of   dry adsorption $(kg.kg^{-1})$

φ is the relative pressure of the water vapor in the gas

$$\varphi = \frac{\text{partial pressure}}{\text{vapor pressure}}$$

Figure 3.1. *Isotherms of adsorption of water vapor (adapted from [KAS 81])*

For numerous gases, adsorption is isothermic (apart from water vapor). Adsorption is absolutely necessary if we need to separate out less than 2% of the volume of the feed, except if there is less than 3 g of impurity per Nm^3 of gas.

3) Liquids

If the temperature falls then q increases. However, in order to decrease the viscosity of the liquid (and therefore trigger material transport), we are led to increase the temperature. Thus, sugar solutions are treated at 80 or 90°C.

Depending on the parameter n of Freundlich's equation, $q = C^{1/n}$:

– adsorption is easy if $0.1 < n < 0.5$;

– adsorption is difficult if $0.5 < n < 1$.

3.2. ("Batch") adsorption from a load

3.2.1. Contact adsorption followed by filtration

This process is valid for liquids. We place 0.1 to 2% mass of adsorbent into the solution in relation to the liquid.

Non-carbonated adsorbents are appropriate for use with non-aqueous solutions (oils). Powdered charcoal can be used for water (with a few ppm of charcoal, the solution needs to be stirred for between 4 and 16 hours).

However, with a pre-layer of between 0.6 and 5 cm, the velocity will be $0.4 \ m.h^{-1}$ for a thickness of 2.5 cm. Contact will last for 3.75 minutes.

The pre-layer is necessary for industrial dry cleaning, sugar solutions and solutions of metal ions, for example.

With resins that are strongly acidic or strongly basic, the exchange lasts for a few minutes. We use 1.2 volumes of solution to each 1 volume of loose resin.

If grime is present (e.g. pectin, ore pulp), even dilute mash cannot be treated in a fixed bed.

In charcoal-treatment tanks, the concentration of the charcoal is low and therefore it poses no danger in terms of corrosion. However, when the concentration is greater than or equal to $60 \ g.L^{-1}$, precautions must be taken.

3.3. Definition of an industrial chromatography column

3.3.1. *Design of the installation*

The principle of the installation is as follows:

– a column with internal baffles;

– a feed device to introduce the solvent or the inert species;

– an injection device for the load to be fractionated;

– a continuous-flow analyzer at the outlet of the column;

– an array of recipients, each of which receives the component corresponding to a given band;

– a faucet mechanism directing the output from the column into the appropriate recipient;

– between two bands, we recover solvent or impure inert material, which is recycled along with the feed.

Certain imperfections may arise in an industrial-scale column:

– the solid particles may be crushed by the weight of the particles above them,

– the temperature may not be uniform over a section of the column;

– the flow of the fluid may not be uniform over a section of the column.

We can remedy the problem of crushing by dividing the solid load into sections, each of which is supported by a metal grill.

The non-uniformities can be dealt with by building in a set of baffles the design of which is described in the [BAD 66] patent which, by now, has lapsed and is in the public domain.

3.3.2. *Number of theoretical plates*

There is a relation which exists between the standard deviation σ_t, which is measured in time, and the standard deviation σ_L, measured in length.

$$\sigma_L = U\sigma_t \text{ where } U = u\left(1 + F\frac{dq}{dc}\right)^{-1} = \frac{u}{1+k'} = \frac{Z}{t_R} \text{ because } u = \frac{Z}{t_0}$$

Because $\sigma_L^2 = HZ$ (see section 2.4.4), the HETP is:

$$H = \frac{\sigma_L^2}{Z} = \frac{\sigma_t^2 U^2}{Z} = \frac{\sigma_t^2 Z}{t_R^2}$$

σ: standard deviation of a concentration band: s or m;

H: height of a theoretical plate: m;

Z: total height of the column: m.

The number of theoretical plates is therefore:

$$N = \frac{Z}{H} = \left(\frac{t_R}{\sigma_t}\right)^2$$

By assigning a particularly important role to component 1, we can write:

$$N = 16\left(\frac{t_{R1} - t_{R2}}{4\bar{\sigma}_t}\right)^2 \times \left(\frac{t_{R1} - t_0}{t_{R1} - t_{R2}}\right)^2 \times \left(\frac{t_{R1}}{t_{R1} - t_0}\right)^2 = 16R_1^2 \times \left(\frac{\alpha}{\alpha - 1}\right)^2 \left(\frac{k_1 + 1}{k_1}\right)^2$$

In order to do so, we simply need to have set:

The resolution:

$$R_1 = \frac{t_{R1} - t_{R2}}{4\bar{\sigma}_t}$$

The separation parameter:

$$\alpha = \frac{t_{R1} - t_0}{t_{R2} - t_0}$$

The capacity parameter:

$$k_1 = \frac{t_{R1} - t_0}{t_0}$$

The length of stay of the fluid phase:

$$t_0 = \frac{Z}{u}$$

u: interstitial velocity of the fluid: $m.s^{-1}$.

Here, we had set: $\bar{\sigma} = \frac{1}{2}(\sigma_1 + \sigma_2)$

The number of theoretical plates in the column is therefore:

$$N = 16R_1^2 \left(\frac{\alpha}{\alpha-1}\right)^2 \left(\frac{k_1+1}{k_1}\right)^2$$

3.3.3. Relationship between the number of plates and the HETP [GLU 64]

Let us set:

N_{IB}: number of plates measured for a brief injection;

H: height equivalent to a theoretical plate (HETP): m;

H_{IB}: HETP measured for a brief injection: m;

t_p: duration of a rectangular injection: s;

t_R: retention time: s.

The [GLU 64] relation is:

$$N_{IB}^{1/2} \left(\frac{t_p}{t_R}\right) = 3.26 \left[\frac{H}{H_{IB}} - 1\right]^{-1/2}$$

3.3.4. *Production rate (liquid chromatography)*

The production rate is defined by:

$$Q = AVc \frac{t_p}{t_{cy}}$$

V: velocity in an empty bed (constant) of the fluid mixture: $m.s^{-1}$;

A: section of the column: m^2;

t_p: duration of the rectangular injection: s;

t_{cy}: duration of a cycle (interval between two injections): s;

c: concentration of the injection: $kmol.m^{-3}$.

Timmins *et al.* [TIM 69] propose the following empirical expression (based on Gluckauf's equation from [GLU 64]):

$$\frac{t_p}{t_{cy}} = \frac{0.4}{R_s}\left(1 - \frac{L_{sep}}{L}\right)^{1/2}$$

L_{sep}: length of the column devoted to separation.

The production rate is, finally:

$$Q_L = \frac{0.4}{R_s} AVc \left(1 - \frac{L_{sep}}{L}\right)^{1/2}$$

Q_L: production rate: $kmol.s^{-1}$

$$L_{sep} = H_{IB} N_{IB}$$

N_{IB}: number of theoretical plates for a brief injection;

H_{IB}: HETP measured for a brief injection: m.

From a different perspective, [KNO 86] propose an approach to maximize production in preparative chromatography, i.e. in industrial chromatography.

3.3.5. *Pressure drop across a fixed bed*

We tend to use Ergun's formula [DUR 99]:

$$\frac{\Delta P}{Z} = 150\frac{\mu V(1-\varepsilon)^2}{d^2\varepsilon^3} + 1.75\frac{\rho V^2}{d}\frac{(1-\varepsilon)}{\varepsilon^3}$$

ΔP: pressure drop: Pa;

Z: height of the bed: m;

μ: viscosity of the fluid: Pa.s;

V: velocity of the fluid in an empty bed: m.s^{-1};

d: diameter of the solid particles: m;

ρ: density of the fluid: kg.m^{-3};

ε: porosity (empty fraction of volume) of the bed: dimensionless.

[FAI 69] gives a table listing the physical properties of the main adsorbent solids.

EXAMPLE.–

A molecular sieve

d = 3.2 mm   Z = 3 m ρ = 2.6 kg.m^{-3}

ε = 0.34 μ = 20.10^{-6} Pa.s   V = 0.1 m.s^{-1}

$$\Delta P = \frac{150\times3\times20.10^{-6}\times0.1\times0.66^2}{3.2^2.10^{-6}\times0.34^3} + \frac{1.75\times3\times2.6\times0.01\times0.66}{0.0032\times0.34^3}$$

$$\Delta P = 974 + 716$$

$$\Delta P = 1690 \text{ Pa} = 0.17 \text{ bar}$$

NOTE.–

This calculation gives a fairly slight pressure drop if the inlet pressure were, say, equal to 1.5 bar abs. Otherwise, we would need to numerically integrate Ergun's equation by setting:

$$dP = f(V)dZ = f\left(\varnothing_M \frac{RT}{P}\right) dZ$$

That is:

$$Z = \int_{P_1}^{P_2} \frac{dP}{f\left(\varnothing_M \dfrac{RT}{P}\right)}$$

$\varnothing_M$: molar flux density

3.3.6. *Adsorption in a fixed bed and breaking time*

Whilst adsorption is taking place, the fluid emerges from the column with a concentration of zero. However, when the adsorption reaches the limit of its capacity, the output concentration begins to rise from that zero value. We set a specification c_1 beyond which adsorption must be halted so as not to obtain an impure fluid. The time corresponding to c_1 can be called the *breaking time because, from that moment on, the specification is no longer respected (it is broken).*

[COL 67] introduced the concept of the unused height in a fixed bed. Let us set:

H_{to}: total height occupied by the adsorbent: m;

H_{th}: theoretical height: m;

G: flux density in mass of the fluid: $kg.m^{-2}.s^{-1}$;

θ_b: breaking time: s;

Y and X: dry humidity of the gas and of the adsorbent: dimensionless.

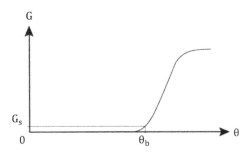

Figure 3.2. *Output concentration*

– Breaking time

The theoretical height is:

$$H_{th} = G\theta_b \frac{\Delta Y}{\Delta X}$$

Y and X are linked by the isotherm:

The unused height is then:

$$H_{in} = H_{to} - H_{th}$$

A pilot installation enables us to determine θ_b and therefore H_{in}

[HUT 73] used three pilot columns of different heights and obtained a bundle of lines showing the breaking time as a function of the height. For each line, there is a corresponding value of the breaking concentration or a value of the liquid flowrate (however, the constant term in the equation of the lines is given by an expression which seems incorrect). This empirical study enabled the author to define an industrial unit.

3.4. Practical use of adsorbents

3.4.1. *Characteristics of charcoal in adsorbents*

The granules range in size between 0.5 and 1 mm (fixed, fluid or mobile bed).

The apparent density of the charcoal is of the order of 320 $kg.m^{-3}$

The active surface of activated charcoal is approximately 1,000 m^2 per gram of charcoal. For good efficiency, the pore diameter needs to be between 0.5 and 5 nm. Certain molecules are too large to penetrate into pores whose diameter is smaller than 0.5 nm.

Charcoal adsorbs hydrophobic compounds (such as solvents).

3.4.2. *Regeneration of charcoal*

The flowrate density of superheated water vapor varies from 0.1 to 0.3 $kg.m^{-2}.s^{-1}$. The heating can also be performed with a coil immersed in the charcoal, or indeed with a double envelope (jacket) for small or very small installations. The vapor is condensed and the solvents are recovered by decantation in appropriate device (a florentine).

Vapor-based regeneration must be avoided if the absorbates polymerize at high temperatures in the presence of humidity. Thus, in a pill-coating factory, a vacuum is created over the charcoal. That vacuum may be up to 1 torr abs. The vapors are recovered in a refrigerated condenser.

For economic reasons, it is common to leave an "ogee" of adsorbate after regeneration.

3.4.3. *Necessary precautions: explosion, hot points (charcoal)*

In the gas at hand, the proportion of flammable gas must always be less than 25% in terms of volume. However, if the adsorption is exothermic, in order to prevent the formation of hot points, the concentration of the flammable compound must be no higher than 1% volume.

The exothermic decomposition of ketones, aldehydes, organic acids and esters is favored by carbon and its impurities. Thus, we see the emergence of hot points if the gas circulation is interrupted. It is crucial to put in place a CO and CO_2 alarm and a water sprinkler system (a deluge!).

3.4.4. Corrosion

Charcoals washed with acid and with water are highly electropositive. The formation of galvanic cells with a metal causes corrosion by cracking. Therefore, we must prevent direct contact with a metal in the presence of an electro-conductive liquid (which is why it is helpful to use ebonite or zinc galvanization), but it is important that the thermal expansion of the coating is the same as that of the coated metal if the regeneration is to be done by heating.

3.4.5. Fixed bed [HEL 52, MER 52]

1) [HEL 52] gives a great deal of practical information about granulated or powdered adsorbents, about the influence of the temperature and about the regeneration of the adsorbent, all in the liquid phase.

2) For dechlorination of water by charcoal, the height Z of the bed is of the order of 0.75 m.

The aim is to evacuate a few ppm of chlorine.

3) For the refining of sugar over activated charcoal, the column height is approximately 6-8 m of animal char in grains.

4) For purification of air loaded with solvents, the bed height varies from 0.3 to 0.9 m of charcoal. The duration of the adsorption operation is around an hour. If a ventilator is used for the air, the pressure drop across the bed is between 0.25 and 0.5 mH_2O. The velocity of air in an empty bed across the bed is 0.3–0.5 $m.s^{-1}$. The recovery yield of a bed is at least 99% of the mass of solvents introduced.

Prior washing of the gas with water eliminates the poisons from the charcoal.

3.4.6. Gas-fluidized mobile bed [CHA 51, MER 52]

The column contains five fluidization plates. Of these:

– the bottom plate is heated by vertical tubes containing a heat-transfer fluid;

– the top plate is cooled by water.

The charcoal passes from one plate to that which is immediately below it, through circular spillways blocked at the base by a horizontal valve supported by a spring. The weight of the solid opens the valve.

The charcoal is brought back to the top by entrainment with the air in an air-lift tube, at the top of which the air is evacuated.

3.4.7. Activated alumina: fixed bed

For the dehydration of gases and liquids, we use anhydrous alumina:

$$Al(OH)_3 \xrightarrow{activ} Al_2O_3 \text{ amorphous and porous (92\%)}$$

Impurities: Na_2O, SiO_2, Fe_2O_3, TiO_2.

The alumina can be reactivated up to a temperature of 315°C (beyond this, the temperature alters the product, and it loses some of its qualities).

The apparatus is heated by electrical resistors or coils (vapor, heat-transfer fluid).

Alumina is used in the vents of oil tanks and other liquid containers.

Porosity 0.75, of which 0.25 is pores.

Apparent density $\rho_a = 830 \, kg.m^{-3}$.

Can dehydrate a gas (2 mg.Nm^{-3}).

Saturation is 25% of its weight in water at ambient T. Above 100°C, adsorption decreases.

0.5 to 1 $m^3.h^{-1}.kg^{-1}$ of alumina is the gaseous flowrate.

Size: 0.65 mm to 1.5 mm.

The bed is fed from top to bottom. However, hot-air regeneration takes place from bottom to top. This being the case, if the regeneration is incomplete, the remaining humidity will be left at the inlet for the humid gas and the vapor will be released more easily.

A calorifuge maintains the heat during reactivation.

Galvanized steel or ebonite, or indeed aluminum or Monel™, offer good corrosion-resistance.

Dehydration can also be performed by:

Al_2O_3

P_2O_5

H_2SO_4

$CaCl_2$

$BaClO_4$ and $Mg\ ClO_4$ (perchlorate).

3.4.8. *Fluidized plates (silica) [ERM 61]*

Silica gel comes in grains of between 75 μm and 500 μm .

The velocity of the gas in an empty bed is 1.3 $m.s^{-1}$. This velocity causes the entrainment of particles finer than 400 μm.

[ERM 61] proposes the following for a dehydration column:

Function	Number of plates
Adsorption	6
Regeneration	5

The author uses McCabe and Thiele's method to simulate the column. He evaluates the heat transfer between the gas and gel, and the equivalence of each plate (equal to 0.8 theoretical plates for a 30 kg.m^{-2} load).

3.4.9. Fixed bed (silica) [JUR 52]

The height H of the bed is 5–6 m. The gas is fed in at a velocity in an empty bed equal to:

$$U = \frac{Q_g}{S} = \frac{0.0135}{60} \times H \times \rho_a$$

ρ_a : apparent density of silica: kg.m^{-3}.

The poisons of silica are:

– water;

– N_2, S and oxygenated compounds;

– dust – particularly with a fixed bed.

NOTE.–

It has become practice to use silica combined with N-octadecylndimethylaminosilane, simply called silica C_{18}.

3.4.10. Zeolites

These minerals crystallize as nanocrystals in the form of a cubic box with windows. The size of these windows is what could be called their aperture, which is measured in nm.

With zeolites, strictly speaking, it is not adsorption on a surface which we see, but capture. Indeed, the captured product comes through the windows into the zeolite molecule, where it remains captive. The van der Waals forces are often responsible for keeping the captured product in place.

The general chemical composition of zeolites is:

$$M_x^{n+} \left[\left(AlO_2 \right)_x \left(SiO_2 \right) \right]_n^{x-} \quad \text{(M is a metal)}$$

The aperture (of the windows) of the zeolite depends on two factors:

a) the types of crystalline shape. There are three which are most common:

A, X, Y

The Y form corresponds to faujasite. Lapis lazuli (a sky-blue stone) is a zeolite containing sulfur;

b) the atomic radius of the metal. The radii of the most widely-used metals are as follows, in decreasing order (in Angström):

K	Ca	Na	Ag
2.36	1.97	1.89	1.44

Broadly speaking, we can say that the aperture of a zeolite varies approximately *in the opposite direction* to the atomic radius of the metal for the same crystalline form.

Let us cite a number of zeolites in increasing order of their apertures (in Angström):

Designation	Metal	Examples of products captured
A3	K^+	H_2O, CO, NH_3
A4	Na^+	H_2S, SO_2, CO_2
A5	Ca^{++}	CHF_3, $(CH_3)_2\,NH$
X9	Na^+	Isoparaffins, $(C_2H_5)_3\,N$

Table 3.1.

Zeolites are essentially used for the following operations:

– catalysis;

– adsorption (drying to a dew point of -40°C for gases);

– ion exchange.

3.4.11. *Resins: nature*

Resins occur in the form of spherules of 1.7 to 3 mm in diameter. They are composed of a polystyrene–divinylbenzene skeleton. A high degree of cross-linking reduces swelling by a liquid, the exchange speed and the capacity.

1) Cationic resins contain sulfonic groups (strongly acidic) or carboxylic groups (weakly acidic). Carboxylic resins have a high exchange capacity and are used to treat large organic ions (such as streptomycin). The zeolite which is a silicate of Al and Na, is a cationic exchanger for a pH near to neutral.

2) Anionic resins contain primary, secondary or tertiary amine radicals (strongly anionic). Strongly-anionic resins capture silicic radicals and anions of salts. Amine resins have a high capacity. They are regenerated by NaOH, NH_4OH or $NaCO_3$.

3.4.12. *Industrial uses of resins*

1) Demineralization of water for:

– high-pressure boiler supply;

– pharmaceutical preparations;

– anywhere that highly-purified water is needed.

2) Recovery of metals: the anodization of aluminum by chromic acid dissolves a certain portion of the aluminum, which can then be recovered by an RH resin which is used to regenerate the chromic acid.

3) Purification of glycerin and formaldehyde.

4) Water softening.

Further information may be found in the article by [HIG 54] and in [HIG 64].

3.4.13. *Longevity of resins*

1) Physical fouling by solid suspensions. This fouling is measured by calcinations.

2) Chemical fouling (molybdenum) which withstands elution. The colloidal precipitation of SiO_2 is also an example of chemical fouling. Sodium dissolves SiO_2, and a 10% H_2SO_4 solution dissolves molybdenum.

3) Destruction of the structure of the resin by oxidation, catalyzed by multivalent metals (Cr, Va, etc.)

In general, the annual loss of resin for a mobile bed is no greater than 10% of the mass used.

3.4.14. *Implementation of resins*

1) In a fixed bed. The bed experiences four operations in turn:

– saturation;

– washing to eliminate fouling;

– regeneration;

– rinsing.

We could, for instance, imagine an array of four columns which are arranged as follows (from the point of view of the connections):

1234   2341   3412   4123   1234

In each column, a gaseous ceiling occupies a volume equal to half or all of the volume of dry resin to absorb the expansion of the bed by swelling of the gel of which the solid particles are made.

2) In a mobile bed [FAI 69], the bed facilitates countercurrent exchange [SOL 67]. The advantage of the mobile bed is that the flowrates are low compared to those of adsorption in a fixed bed. The valves do not work on the resin, and are operated by the injection of liquid (the author's Figure 13).

A 1-meter-high bed is sufficient.

However, the range of acceptable flowrates is limited.

3) Fluidized bed on two plates, also containing a regeneration zone outside of the column.

4) Pulsed jig.

3.5. Non-isothermal adsorption

3.5.1. *A few notations to begin with*

Let us set:

α: internal porosity of the solid particles;

ε: porosity of the bed (external to the particles);

a: volumetric surface area of a particle: $m^2.m^{-3}$

$$a = 6/d_p$$

d_p: diameter of a spherical particle (spherule): m.

The product α a is the area of the pores available to the fluid at the surface of the spherules, expressed per $1 \ m^3$ of spherules.

φ: concentration of the adsorbate in the solid: kmole per m^3 of spherules.

3.5.2. *A proposal for the diffusivity in adsorption in a porous medium*

If, initially, we suppose the absence of adsorbate in the spherule, and the concentration of adsorbate outside of the particles is c_A, when the adsorbate enters into the pores, it forces out the inert species which filled them previously. That volume of inert species is equal to the volume of adsorbate found in the fluid phase, to the exclusion of the adsorbate fixed to the wall of the pore.

Let us set:

N_{ad}: flux density of adsorbate fixed to the walls: $kmol.m^{-2}.s^{-1}$;

N_{av}: flux density of adsorbate filling the pores in the fluid phase: $kmol.m^{-2}.s^{-1}$;

N_I: flux density of the inert species evicted: $kmol.m^{-2}.s^{-1}$;

q: overall concentration (walls and volume of the pores) per m^3 of solid: $kmole.m^{-3}$.

This flux is impeded by a flux of inert species in the opposite direction:

$$N_I = -N_{av}$$

Finally, the Maxwell–Stefan equation for the adsorbate is written (see section 4.2.1 of [DUR 16a]):

$$-q\frac{dLn\gamma q}{dz} = \frac{1}{D}(-N_I x_a + N_A x_I)$$

$$N_I = -N_{av} \qquad N_A = N_{av} + N_{ad} \qquad x_I = 1 - x_A$$

Thus:

$$-q\frac{dLn\gamma q}{dz} = \frac{1}{D_{ad}}\left[N_{av} + N_{ad}(1 - x_A)\right]$$

If the flux density N_{av} is negligible in comparison to the term $N_{ad}(1-x_a)$, we obtain the simple equation:

$$-q\frac{dLn\gamma q}{dz} = \frac{1}{D_{ad}}N_{ad}\left(1 - \frac{c_A}{c_T}\right)$$

This classic equation differs little from Fick's equation.

If N_{av} needs to be conserved, we cannot change anything, but there is a relation which exists between N_{ad} and N_{av}.

Let us set:

r_p: radius of the pore: m;

L: length of a pore: m;

$q*(c)$: concentration per m^3 of spherules at equilibrium with c: kmol.m^{-3}.

The ratio of the numbers of moles involved is:

$$\frac{M_{av}}{M_{ad}} = \frac{\pi r_p^2 c_A L}{2\pi r_p q_s^*(c)L} = \frac{r_p}{2q_s^*(c_A)} = \emptyset$$

However:

$$\frac{N_{av}}{N_{ad}} = \frac{M_{av}}{M_{ad}} = \emptyset \text{ hence } N_{av} = \emptyset N_{ad}$$

The Maxwell–Stefan equation becomes:

$$-q\frac{d\ln\gamma q}{dz} = \frac{1}{D_{ad}}N_{ad}\left[\left(1-\frac{c_A}{c_T}\right)+\emptyset\right] = \frac{1}{D_e}N_{ad}$$

where, in a pore:

$$D_e = \frac{D_a}{\emptyset+1-\dfrac{c_A}{c_T}}$$

D_e is the effective diffusivity and D_a is the diffusivity of the adsorbent in the fluid mixture.

The diffusivity in a spherule is:

$$D_p = \frac{\alpha D_a}{t^2\left(\emptyset+1-\dfrac{c_A}{c_T}\right)}$$

c_T: total concentration in the fluid (including inert species);

α: empty fraction in a spherule;

t^2: tortuosity of the pores (usually between 2 and 3, or sometimes 4).

Finally, the diffusion equation is:

$$\frac{1}{4\pi r^2}\frac{\partial}{\partial r}\left[4\pi r^2 D_p \frac{\partial q}{\partial r}\right]=\frac{\partial q}{\partial t}$$

By integrating this equation, we obtain the concentration field in the spherule.

3.5.3. *Local heat balance in a spherule*

The specific heat capacity of the pores is the sum of five terms:

1) Those pores not reached have the following specific heat capacity:

$$\left(1-\frac{q}{q_\infty}\right)\alpha\,C_{fO}$$

α: porosity of the spherule;

q: concentration of the adsorbate in the spherule: kmol per m³ of particles;

C_{GO}: specific heat capacity of the pre-existing fluid: $J.m^{-3}.°C^{-1}$.

2) The adsorbate fixed has the following capacity:

$$qC_{AL}$$

C_{AL}: specific heat capacity of the adsorbate in the liquid state: $J.kmol^{-1}.°C^{-1}$.

3) The fluid fed in which has penetrated into the pores has the following capacity:

$$\alpha\frac{q}{q_\infty}\,C_{f1}$$

C_{f1}: specific heat capacity of the fluid penetrating into the pores: $J.m^{-3}.°C^{-1}$.

4) In total, the specific heat capacity of the contents of the pores is:

$$M_p(q) = \left(1 - \frac{q}{q_\infty}\right)\alpha\, C_{f0} + qC_{AL} + \frac{q}{q_\infty}\alpha\, C_{f1}$$

5) The specific heat capacity of the solid of a particle is:

$$(1 - \alpha)\rho_s\, C_s$$

ρ_s: true density of the solid: $kg.m^{-3}$;

C_s: specific heat capacity of the solid: $J.kg^{-1}.°C^{-1}$.

6) The specific heat capacity of a m^3 of spherule is:

$$M_s(q) = M_p(q) + (1 - \alpha)\rho_s\, C_s \quad (J.m^{-3}.°C^{-1})$$

The local increase in heat energy is, at radial distance r (measured in $J.m^{-3}.°C^{-1}$):

$$\frac{1}{4\pi r^2}\frac{\partial}{\partial r}\left[4\pi r^2\left(\lambda\frac{\partial T}{\partial r} + D_p\Delta\frac{\partial q}{\partial r}\right)\right] = M_s(q)\frac{dT}{d\tau}$$

λ: heat conductivity: $J.s^{-1}.m^{-1}.°C^{-1}$;

T: temperature within the spherule: °C;

D_p: diffusivity of the adsorbate in the spherule: $m^2.s^{-1}$;

Δ: isosteric heat of adsorption: $J.kmol^{-1}$;

q: concentration of adsorbate: kmol per m^3 of spherule;

τ: time: s.

3.5.4. Heat balance of the fluid

The heat received by the fluid per second and per m^3 of the column is the sum of four terms (expressed in $W.m^{-3}$).

1) A term representing the transport by the fluid phase

$$\varepsilon u C_{f1} c_T \frac{dT}{dz}$$

u: interstitial velocity: $m.s^{-1}$;

ε : empty fraction of the column outside of the solid;

c_T : total concentration of the fluid (including the inert species): $kmol.m^{-3}$;

z: depth into the column from the inlet: m;

C_{f1} : molar specific heat capacity of the fluid: $J.kmol^{-1}.{}^{\circ}C^{-1}$.

2) A term of exchange between fluid and solid

$$a(1-\varepsilon)h_f(T_s - T) \qquad\qquad (W.m^{-3})$$

T_{ss}: surface temperature of spherules: $^{\circ}C$;

hf: coefficient of fluid–solid heat convection: $W.m^{-2}.{}^{\circ}C^{-1}$;

a: volumetric surface area of the spherules: $m^2.m^{-3}$.

3) A term of exchange with the wall

$$\frac{\pi D_C}{A} h_{wall}(T - T_{wall}) \qquad\qquad (W.m^{-3})$$

D_C : diameter of the column: m;

A: section of the column: m^2;

h_{wall} : coefficient of fluid–wall transfer: $W.m^{-2}.{}^{\circ}C^{-1}$.

4) A term of local accumulation

$$\varepsilon C_{f1} c_T \frac{dT}{d\tau}$$

τ : time: s.

5) The heat balance equation of the fluid is therefore written as follows:

$$\varepsilon u C_{fl} c_T \left(\frac{\partial T}{\partial \tau} + u \frac{\partial T}{\partial z} \right) = a(1-\varepsilon) h_f \left(T_{ss} - T \right) + \pi D_c h_{wall} \left(T_{wall} - T \right)$$

3.5.5. *Equations of fluid–solid exchanges*

These exchanges (material and heat) take place at the surface of the particles ($r = r_p$, the radius of a spherical particle):

mass $$D_p \left. \frac{\partial q}{\partial r} \right|_{r_p} = k_f \left(C_A - C_A^* \right)$$

heat $$\lambda \left. \frac{\partial T}{\partial r} \right|_{r_p} = h_{solid} \left(T - T_{ss} \right)$$

C_A^* : concentration of the solute in the fluid at equilibrium with the concentration q of the solid: kmol.m^{-3};

T_{ss} : surface temperature of the solid: °C;

T: temperature of the fluid: °C;

k_f : material transfer coefficient across the fluid film surrounding the particles: m.s^{-1};

h_{solid} :   heat transfer coefficient across the fluid film surrounding the particles: W.m^{-2}.°C^{-1}.

NOTE.–

[MEY 67] proposed a numerical method for solving the above equations – admittedly with somewhat different notations.

3.5.6. *Intra-corpuscle diffusivity*

Ruthven *et al.* [RUT 80] used a microbalance to measure the evolution of the adsorbed quantity in the presence of a gaseous atmosphere in motion. The equations of the problem were:

$$\frac{\partial q}{\partial t} = D_s \frac{1}{r^2} \frac{\partial}{\partial^2} \left(r^2 \frac{\partial q}{\partial r} \right)$$

$$\bar{q} = \frac{3}{4\pi r^3} \int_0^R q\left(4\pi r^2\right) dr = 3 \int_0^R qr^2 \, dr$$

q: concentration: kg per m^3 of solid;

$\bar{q}$: mean concentration: $kg.m^{-3}$;

r: radial distance: m;

t: time: s;

R: radius of a particle.

The heat balance of the solid phase was:

$$\rho C_p \frac{dT}{dt} = \Delta H \frac{\partial \bar{q}}{\partial t} + ha\left(T_a - T\right)$$

T: temperature of the solid (supposed to be uniform in a particle): °C;

T_a : temperature of the gaseous atmosphere;

h: gas–solid heat-transfer coefficient: $W.m^{-2}.°C^{-1}$;

a: volumetric surface area of the solid: $m^2.m^{-3}$;

ΔH: isosteric heat (the opposite of the heat of adsorption): $J.kg^{-1}$;

ρ: density of the solid phase: $kg.m^{-3}$;

C_p: specific heat capacity of the solid phase: $J.kg^{-1}.°C^{-1}$.

The influence of the temperature on the isotherm is:

$$q_s - q_0 = \frac{\partial q^*}{\partial T}(T - T_a)$$

q_s: concentration at the surface of the granules: $kg.m^{-3}$.

The authors integrate these equations and determine D_s so that the calculation corresponds to the experience of reality. The diffusivity D_s does not depend on the temperature within a 40°C range.

3.5.7. Lee and Cummings' empirical method [LEE 66]

For Lee and Cummings, the adsorbent used is silica gel in spherules whose diameter is 3.9 mm. Those authors use an empirical relation to express a certain parameter α as a function of the relative humidity of the incoming gas. The parameter α is defined by:

$$\alpha = \frac{t_s - \tilde{t}_b}{t_s - t_b}$$

t_s is the saturation time defined by the authors' equation;

$\tilde{t}_b$ is the breaking time for non-isothermal adsorption;

t_b is the breaking time for adsorption supposed to be isothermal.

3.6. Particular cases in chromatography

3.6.1. Gradient elution

A so-called "*gradient* elution" is obtained by changing the affinity of the eluent for the adsorbent. To do so, we add, to the elution liquid, a compound called a *modulator*, whose concentration will vary over time in a linear fashion.

This technique may be useful in preparative chromatography, where the concentrations are high. Indeed, in this case, the solutes may react with one another. This is what happens in so-called "overloaded" chromatography. Whilst, for low concentrations, the isotherms of solutes may be linear and independent of one another, this is often no longer the case in overloaded chromatography.

The concentration of the modulator is given by:

$$\varphi(t,\ z) = \varphi_0 + \beta\left(t - \frac{z}{u_0} - t_{inj}\right)$$

t_{inj}: duration of the rectangular injection: s;

u_0: interstitial velocity of the liquid: $m.s^{-1}$.

The parameter β characterizes the concentration gradient of the modulator.

Suppose that a solute obeys a Langmuirian isotherm:

$$q_i = \frac{a_i c_i}{1 + \sum_j \rho_j c_j}$$

In the presence of the modulator, the equation of the isotherm becomes:

$$q_i = \frac{a_i \exp(-S_i \varphi) c_i}{1 + \sum_j \rho_j \exp(-S_i \varphi)}$$

The presence of the modulator may cause the inversion of the selectivity between solutes i and k. Thus, the selectivity $\dfrac{q_i c_k}{c_i q_k}$ crosses the value 1.

[ANT 89] give a detailed description of the results of gradient chromatography. While linear chromatography yields good results without a modulator, the resolution of the peaks and bands is improved by the modulator in overloaded chromatography. [GHR 88] propose strategies to use the gradient method.

[GIB 86] proposes an approach to switch from the pilot unit to the industrial in the case of gradient chromatography.

3.6.2. Reversed-phase chromatography (RPC)

A polar liquid such as water and a hydrocarbon are not miscible, and therefore a mixture of them contains two phases. Consequently, it is almost impossible to fix a hydrocarbon onto silica, which is polar. This is no longer the case if we graft an alkyl radical to the surface of the silica – the radical then attracts the hydrocarbons by van der Waals attraction.

A classic way of working is to fix an octadecyl radical (in C_{18}) to the silica and to use an eluent which is a mixture of methanol and acetonitrile. Then, because there is no organic species in the composition of the eluent, the only cause of fixation of the solute to the silica is the presence of the octadecyl radical.

Thus, the principle of RPC is based on a hydrophobic effect, and RPC is successfully used for the separation of biopolymers. Horvath *et al.* [HOR 76] describe a use of this method in detail.

3.6.3. Various questions

1) Protein aggregation [WHI 91] in chromatography

If the aggregation is rapid, it is significant. A single peak then encapsulates the peaks of different polymers. Aggregation is a reaction mechanism which we shall not discuss here.

2) "Fingering" [NOR 96] in liquid adsorption

When a viscous liquid is displaced by a much less viscous liquid that is miscible with the first, the less viscous forms "fingers" which push into the more viscous liquid.

NOTE.–

Kaczmarski *et al.* [KAC 02] applied a model taking account of internal diffusion by application of the Maxwell–Stefan law, using the Tóth multiple isotherm (i.e. one pertaining to multiple components), taking account of axial dispersion and accepting that local equilibrium is reached.

3.7. Practice of ion exchange

3.7.1. *Practical regeneration procedures*

1) Simple bed:

– thinning of the bed and evacuation of the deposits by circulation of water from bottom to top, which causes the fluidization and expansion of the bed;

– regeneration from top to bottom;

– slow and then quick rinsing from top to bottom.

2) Mixed bed:

– fluidization and expansion, and particularly separation of the two resins, with the cationic resin (which is heavier) being at the bottom;

– regeneration of the anionic resin with recovery of the regeneration solution of caustic soda between the two layers;

– rinsing of the anionic resin;

– regeneration de la cationic resin;

– rinsing of the cationic resin;

– at least 3 injections of compressed air for 4–5 minutes each so as to mix the two active materials.

3) Percolation rates:

Both in the active phase and in the regeneration phase, the velocity in an empty bed (over an empty section) of the liquid is situated between 8 and

25 m.h^{-1}. Note that this range of velocity is also that of the liquids in a differential liquid–liquid extractor.

A velocity of over 25 m.h^{-1} would cause a major pressure drop and an excessive settling of the bed, which would be attended by the wearing out and fracture of the solid particles, the debris from which would be a cause of clogging of the bed.

3.7.2. Swelling of resins

After immersion in water, the grains of resin generally exhibit swelling due either to the osmotic pressure or to the solvation of free ions which have moved into the resin. However, highly-crosslinked resins, whose skeleton is less "elastic", are least vulnerable to swelling.

This swelling can, in extreme situations, cause the apparition of overpressure of the order of 100 bar on the wall of the column. The probability of occurrence of that overpressure increases with the H/D (height/diameter) ratio of the resin bed. Indeed, in divided-solid mechanics, we can show the following approximate result:

– If H/D > 2, the bed is clogged and there is overpressure.

– If H/D < 2, the resin can slide along the wall, and the expansion of the bed takes place freely in the axial direction.

3.7.3. *Various processes and possible combinations*

We shall now describe a number of setups:

1) Remember that the permutation between alkali and alkali-earth ions is done with a natural clay, which fixes the calcium ions from a solution and relinquishes its own sodium ions into the liquid. This is the old technique of softening.

2) The appearance of synthetic resins and the technique for regenerating them facilitated more economical softening, by the use of a weakly acid cationic resin regenerated with NaCl salt. All salts in the solution in question have become sodium salts.

3) Combining, with the above treatment, a column of strongly basic anionic resin, also regenerated using NaCl, which replaces all the strongly- and weakly-acidic anions with Cl ions in the solution. The solution obtained is then a solution of NaCl salt.

4) Total demineralization using two strong resins in order:

– one strongly-acid cationic resin regenerated with HCl;

– the other strongly basic anionic resin regenerated using sodium hydroxide NaOH.

All the cations are replaced by H^+ and all the anions by OH^-. These ions combine and we obtain pure water which has a resistivity of around $0.1 \ M\Omega.cm^{-1}$.

5) The demineralization will only be partial if the basic resin is weak, because the silicic ions SiO_3^{2-} and carbonic ions CO_3^{2-} will not be retained. The resin will retain only the strong anions: SO_4^{2-}, NO_3^-, Cl^-. Regeneration is performed, as in the case of total demineralization (in step 4) with HCl and NaOH. A thermal degasser is needed to evacuate the CO_2 resulting from the decarbonation by the strongly-acidic resin. Indeed, the treatment applied after a weakly-basic resin would not retain the CO_3^{2-} or HCO_3^- ions.

3.7.4. *Obtaining very pure water by strong demineralization*

If the total demineralization, as described in 4), is performed in a mixed bed, the anions and cations simultaneously undergo attraction from both types of resins, which improves the efficiency of the operation and reduces leakage (in particular, the leakage of CO_2 and SiO_2) and the resistivity reaches $0.5 \ M\Omega.cm^{-1}$.

The device can be supplemented with a cationic column upstream or downstream, and the resistivity then reaches up to $1 \ M\Omega.cm^{-1}$.

We can also split the process into two steps. Two main columns – one cationic, the other anionic – are followed by two finishing columns. The resistivity then reaches $10 \ M\Omega.cm^{-1}$ and, if we insert a mixed bed between the main columns and the finishing columns, we obtain $15 \ M\Omega.cm^{-1}$, which is close to the limit of $27 \ M\Omega.cm^{-1}$ – the resistivity of chemically-pure water.

NOTES.–

Simply softened water has:

– no hardness: TH = 0°f;

– unchanged salinity (in °f).

Simply decarbonated water has:

– unchanged hardness and salinity (expressed in °f);

– a lower pH;

– a greatly-decreased complete alkalinity titration.

3.7.5. *A few applications for ion exchange*

1) Recovery, on sulfonic resins, of chromium from galvanoplasty baths. The dichromate ion $Cr_2O_7^{--}$, in which the chromium is trivalent, is non-toxic, whereas the chromate ion CrO_4^{--}, where the chromium is hexavalent, is toxic.

2) Removal of excess iron for regeneration of chrome plating baths. The resin is regenerated using acid.

3) Total demineralization of residual liquors (in particular the wash liquors from the process of galvanoplasty). Thus, the water can be recycled. In fact, recovery of the solutes in this way is not at all economical, because we may be dealing with the cyanide ion CN^- (which is a deadly poison) or the chromate ion CrO_4^{--} (which is toxic), in which case, economy is not a major concern.

4) Refining of wine by transforming the double salt potassium sodium tartrate into simple sodium tartrate. It is also possible to eliminate iron and copper from wine in this way.

5) Deacidification of sugary juices to prevent precipitation.

Theory of Drying Theory of Imbibition and Drainage. The Phenomenon of Hysteresis

4.1. Properties of products

4.1.1. *What is drying?*

To dry a product is to rid it of the water or solvent which it contains by vaporizing that water or solvent. To do so, we apply heat by radiation, by conduction or indeed (as we shall see here), by convection with a hot gas.

4.1.2. *Phase law*

Let us apply Gibbs' law of variance to a system composed of a wet product in contact with a wet gas. The number of components is c and is equal to three – namely: the dry product, water and the dry gas. In this process, the pressure is constant, and less than or equal to the atmospheric pressure.

The variance is therefore:

$$v = c + 1 - \varphi = 4 - \varphi$$

In the case of crystals coated with saturated parent liquor (which is commonly found in the chemical industry), there are three phases: the solid, the parent liquor and the gaseous phase. The system is monovariant and is fixed by the temperature.

If we now consider a food product, the water is then bonded to the product, and we are now dealing with only two phases: the wet product and the gaseous phase. The system is bivariant and is determined by the temperature and humidity of the product at equilibrium with the humidity of the gas.

4.1.3. *Wet crystals*

To comprehend the drying rate, we shall see that it is necessary to know the vapor pressure of the saturated parent liquor. This value can, of course, be measured, but it can also be calculated, in either of two different ways:

1) Activity-based method

The vapor pressure of the liquor $\pi_p(t)$ is:

$$\pi_p(t) = a_E \pi(t_p)$$

t_p: temperature of the liquor;

a_E: activity of water;

$\pi(t_p)$: vapor pressure of pure water at t_p.

The activity of the water is given by:

$$a_E = \gamma_E x_E(t_p)$$

$x_E(t_p)$: molar fraction of the water in the saturated liquor at t_p;

γ_E : activity coefficient of the water for that molar fraction.

Readers can refer to Chapter 3 of [DUR 16a] for how to calculate γ_E.

2) Boiling-retardation method

Prior to drying, we generally see crystallization, owing to the concentration of the parent liquor, which is usually achieved by vaporization.

At atmospheric pressure, water boils at 100°C. If we dissolve a component in it, the solution boils at a higher temperature, and the excess temperature over 100°C is known as the boiling retardation R_E. We find that R_E varies little with the pressure at which the boiling takes place. The concept of boiling retardation is very widely used by specialists in the field of evaporation.

If water boils at the temperature t_{EB} at atmospheric pressure P_A, we can write:

$$P_A = \pi(t_{EB})$$

$\pi(t)$: saturating vapor pressure of water: Pa.

If the parent liquor boils at the temperature $t_{Boil} + R_E$ at atmospheric pressure P_A, then by definition of the activity a_E of the water in the liquor, we can write:

$$P_A = a_E \, \pi(t_{EB} + R_E)$$

From these two equations, it stems that:

$$a_E = \pi(t_{EB}) / \pi(t_{EB} + R_E)$$

4.1.4. Food products

Food products are characterized by a relation between the humidity X of the product (in the dry state) and the activity of the water a_E.

At a given temperature t_P, this relation is known as a sorption isotherm. The most widely-used expression is the GAB (Guggenheim-Anderson-de Boer) isotherm.

$$X = \frac{X_m CKa_E}{(1 - Ka_E)(1 - Ka_E + CKa_E)}$$

X_m: "monolayer humidity" in the sense of the BET theory;

C: empirical constant that is a function of the temperature

$$C(T) = C' \exp\left(\frac{E_C}{RT}\right)$$

E_C is of the order of 10,000 to 30,000 kJ.kmol^{-1};

C': empirical constant;

T: absolute temperature: K;

R: ideal gas constant: 8314 J.kmol^{-1}.K^{-1};

K: empirical constant dependent on the temperature

$$K(T) = K' \exp\left(\frac{E_K}{RT}\right)$$

E_K may be positive or negative;

K': empirical constant.

Note that, at a given temperature, the GAB isotherm is defined by three empirical constants: X_m, C and K.

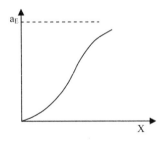

Figure 4.1. *GAB isotherm*

The results are correct for values of a_E less than 0.9, which is ample from the point of view of the drying of food products.

It is possible to express a_E as a function of X by the relation of Figure 4.1.

4.1.5. *Excess latent heat (also called the bond heat)*

The excess latent heat is the heat L_L which needs to be added to the latent heat of evaporation of pure water to extract it from a product by vaporization. Let us distinguish the case of crystals and that of the food products.

1) Crystals

In thermodynamics, it is established that:

$$L_L = M_E RT^2 x_E (t_P) \left(\frac{\partial Ln\gamma_E}{\partial T} \right) \qquad (J.kg^{-1})$$

x_E: molar fraction of water in the parent liquor;

M_E: molar mass of water: 18 kg/kmol.

2) Food products

For these products, it is customary to call the excess heat the bond heat L_L. The theoretical approach which gives us the expression of the GAB isotherm implies that:

For $0 < X \ll X_m$ $L_L = M_E E_c$ bound water

For $X \gg X_m$ $L_L = -M_E E_K$ free water

Between these two extremes, L_L varies continuously, as indicated by Figure 4.2.

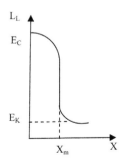

Figure 4.2. *Variation of bond heat with humidity*

To express the total heat of vaporization L – i.e. the sum of the heat of vaporization L_v of the pure water and the bond heat L_L – we merely need to apply Clapeyron's equation. Indeed, the approach leading to the justification of this equation is the same whether we are dealing with the vaporization of a molecule of water from pure water at equilibrium with its vapor, or vaporization of the same molecule from a solution (liquid or solid) at equilibrium with the gaseous phase.

Hence:

$$L = RT^2 \left[\frac{\partial}{\partial T} \{a_e \pi(T)\} \right]_X \qquad \text{(see 1.8.3 vol. 1)}$$

$\pi(T)$ is the vapor pressure of pure water.

By solution of a second-degree equation, the GAB isotherm gives us:

$$a_e = f(C,K,X) = a_e(X,T)$$

4.1.6. Diffusivity of water in plants

In all cases, the diffusivity varies as a function of the temperature in accordance with an Arrhenius-type law:

$$D = D_o \exp\left(-\frac{E}{RT} \right)$$

E is the activation energy for diffusion, and is between 30,000 and 60,000 kJ/kmol.

The variation of D as a function of the humidity is more complicated and depends on the volumetric fractions occupied by:

– the intercellular gas (air + water vapor);

– the cell walls (essentially cellulose);

– the vacuoles (sugary solutions);

– the cytoplasm (proteins and starch).

Furthermore – particularly in wood – the diffusivity along the fibers is twice as great as the transversal diffusivity. In addition, the diffusivity depends on the humidity, and may reach a maximum.

In the case of grains, the results are simpler, and we can agree that the diffusivity remains constant throughout the drying process. Put differently, the analytical solutions to Fick's equation would seem to be justified.

Table 4.1 gives some approximate values of the diffusivity of water in plant materials.

Material	Diffusivity D $(10^{-10}\ m^2/s)$
Wood	10 to 100
Potato	1 to 10
Soya, wheat, cotton	0.1 to 1

Table 4.1. *Diffusivity of water*

We know that, in wheat silos, air is periodically blown in to evacuate the humidity and counteract the heat buildup due to any fermentation.

It must be possible to calculate the evolution of the humidity of wheat during its storage, and the method used draws inspiration from that which is employed in adiabatic adsorption.

With regard to the drying of wood, the practical guide published by the *Centre Technique du Bois* provides useful data.

In particular, according to the authors of the aforecited document, the hygroscopic equilibrium of wood is, with a few rare exceptions, practically not influenced by the variety of tree, and the guide gives a universal equilibrium diagram which expresses the variations in relative humidity of any ideal gas at equilibrium with wood as a function of the temperature and humidity of wood (that relative humidity is identical to the activity of the water in the material in question).

In Melcion and Hari's edited work [MEL 03, p. 334] readers will find a curve showing the diffusivity of water in various agri-food products. That curve was computed by [BRU 80].

4.2. General methods for calculations on dryers

4.2.1. *Material balance*

Whether we are dealing with a continuous-flow or discontinuous-flow dryer, we simply need to write that the water lost by the product has been gained by the gas.

$$G\Delta Y + P\Delta X = 0$$

G: mass flowrate or mass of gas engaged: $kg.s^{-1}$ or kg;

P: mass flowrate or mass of product treated: $kg.s^{-1}$ or kg;

X and Y are the dry humidities of the product and the gas (kg/kg).

The heat balance is trickier to establish.

4.2.2. *Expression of enthalpies*

The temperatures will be evaluated in °C.

1) Dry product

The enthalpy $I_P(t)$ of the dry product is:

$$I_P(t) = C_P t$$

C_P: specific heat capacity per mass: $J.kg^{-1}.°C^{-1}$.

2) Pure liquid water

$$I_E(t) = C_E t$$

C_E: specific heat capacity per mass of liquid water: $J.kg^{-1}.°C^{-1}$.

3) Water vapor

$$I_V(t) = C_E t + L(t) = C_V t + L(0) = C_V t + L_0$$

L: latent heat of vaporization of pure water: $J.kg^{-1}$;

C_V: specific heat capacity per mass of the vapor: $J.kg^{-1}.°C^{-1}$.

4) Bound water

We can obtain vapor by evaporating either pure water or bound water:

$$I_V(t) = C_E t + L_E(t) = I_{EL}(t) + L_E(t) + \int_0^X L_L(X)dX$$

From this expression, we derive the enthalpy $I_{EL}(t)$ of the bound water:

$$I_{EL}(t) = C_E t - \int_0^X L_L(X)dX$$

5) Wet product

$$\overline{I_P(t)} = I_P(t) + X\,I_{EL}(t)$$

$$\overline{I_P(t)} = (C_P + XC_E)t - \int_0^X L_L(X)dX$$

This enthalpy is expressed in relation to 1 kg of dry product.

6) Dry gas

$$I_{GS}(t) = C_G t$$

C_G: gravimetric specific heat capacity of the dry gas: $J.kg^{-1}.°C^{-1}$.

7) Wet gas

$$I_G(t) = (C_G + YC_V)t + YL_L$$

This enthalpy is expressed in relation to 1 kg of dry gas.

4.2.3. *Theoretical heat balance*

We simply need to write that the heat exchanged has only been exchanged between the gas and the product:

$$G\Delta I_G + P\Delta I_P = 0$$

For a continuous-flow dryer, let indices 0 and 1 represent the inlet and outlet of the gas and the product or, in the case of a charge dryer, the initial and final states of the gas and product.

The increase in enthalpy of the product is written as follows:

$$\Delta I_P = I_{P1} - I_{P0} = (C_P + X_1 C_E) t_{P1} - (C_P + X_0 C_E) t_{P0} - \int_{X_0}^{X_1} L(X) dX$$

When we add and subtract $X_1 C_E t_{P0}$, we find:

$$\Delta I_P = (C_P + X_1 C_E)(t_{P1} - t_{P0}) + C_E t_{P0}(X_1 - X_0) - \int_{X_0}^{X_1} L(X) dX$$

The increase in the enthalpy of the gas is written:

$$\Delta I_G = (C_G + Y_1 C_V) t_{G1} - (C_G + Y_0 C_V) t_{G0} + L_0(Y_1 - Y_0)$$

By adding and subtracting $Y_0 C_V t_{G1}$, we obtain:

$$\Delta I_G = (C_G + Y_0 C_V)(t_{G1} - t_{G0}) + (L_0 + C_V t_{G1})(Y_1 - Y_0)$$

However, the quantity of water evaporated is:

$$\Delta E = G(Y_1 - Y_0) = -P(X_1 - X_0) > 0$$

Thus, we have the theoretical balance:

$$G(C_G + Y_0 C_V)(t_{G0} - t_{G1}) = P(C_P + X_1 C_E)(t_{P1} - t_{P0}) + \\ \Delta E(L_0 + C_V t_{G1} - C_E t_{P0} + \overline{L_L})$$

where:

$$\overline{L_L} \, \Delta E = \int_{X_0}^{X_1} L(X) dX$$

IN OTHER WORDS.–

The cooling of the air introduced serves to heat the product, so that it passes from t_{p0} to t_{p1}, and to bring the evaporated water from the initial state (bound liquid at t_{p0}) to the final state (vapor at t_{G1}).

4.2.4. Practical heat balance

Let us write that the incoming heat is equal to the outgoing heat:

$I_{G0}G$	+	$I_{P0}P$	+	Q_H	+	W_v	+	W_s
Gas feed		Product feed		Internal heater		Ventilation power		Transport power

	=	$I_{G1}G$	+	$I_{P1}P$	+	Q_L
		gas outlet		product outlet		heat losses

The energy cost of the drying process sometimes accounts for over 25% of the total cost (including amortization).

The above balance can be written differently if, for the enthalpies, we set:

$$\Delta I_G = I_{G1} - I_{G0} \quad \text{and} \quad \Delta I_P = I_{P1} - I_{P0}$$

$$G \Delta I_G + P \Delta I_P = Q_H + W_v + W_s - Q_L$$

4.2.5. Hypothesis of uniform temperature of the crystals

In view of the usual size of crystals, it is easy to verify that their temperature becomes homogeneous after a very short period of time.

For a sphere of diameter d, the minimum value of the heat transfer coefficient α is (Re = 0):

$$\alpha = \frac{2\lambda_G}{d}$$

The Biot number is at least:

$$Bi = \frac{\alpha d}{\lambda_G} \# 2$$

Let us calculate the Fourier number:

$$Fo = \frac{\lambda_p \tau}{\rho C d^2}$$

where:

$\lambda_p = 2 \text{ W.m}^{-1}.^{\circ}C^{-1}$ $\tau = 0.1 \text{ s}$ $\rho = 1000 \text{ kg.m}^{-3}$

$C = 2\,000 \text{ J.kg}^{-1}.^{\circ}C^{-1}$ $d = 2\times10^{-4} \text{ m}$

$$Fo = \frac{2\times0,1}{1000\times2000\times4.10^{-8}} = 2.5$$

In Figure 34.3 of Perry [PER 73], we can see that, at the center of the sphere:

$$\frac{t_s - t}{t_s - t_o} < 10^{-3}$$

t_s: surface temperature of the solid: °C;

t_o: initial temperature of the sphere: °C;

t: temperature at the center of the sphere: °C.

Thus, for the operation of drying, we assimilate the crystals to spheres of homogeneous temperature, covered with a layer of saturated parent liquor.

4.2.6. *Material transfer from the solid to the gas*

We can express the flux density of solvent transferred from the product to the gas across the film of inert species enveloping the solid, as we do with the condensation of a vapor in the presence of an inert species. We merely need to integrate the Maxwell–Stefan equation.

If the thickness of the limiting film is δ, the result is as follows:

$$N_E = \frac{C_T D}{\delta} Ln\left[\frac{1-y_G}{1-y_P}\right] = C_T \beta Ln\left[\frac{1-y_G}{1-y_P}\right]$$

C_T: total concentration of the gaseous phase: $kmol.m^{-3}$;

D: diffusivity of the solvent in the gas: $m^2.s^{-1}$;

y_G: molar fraction of solvent in the gas.

If π_p is the vapor pressure of the parent liquor and P_T the total pressure:

$$y_P = \frac{a_E \pi_P}{P_T}$$

a_E: activity of water.

The humidity contents are generally low enough for us to be able to write:

$$\frac{1-y_G}{1-y_P} = 1 + \frac{y_P - y_G}{1-y_P} \# 1 + (y_P - y_G)$$

The transfer flux density is therefore:

$$N_E = c_T \beta(y_P - y_G) \qquad \left(kmol.m^{-2}.s^{-1}\right)$$

and if we think in terms of mass:

$$\phi_E = M_E c_T \beta(y_p - y_g) \qquad \left(kg.m^{-2}.s^{-1}\right)$$

c_T: total molar concentration: $kmol.m^{-3}$;

M_E: molar mass of water: $kg.kmol^{-1}$.

NOTE.–

The molar fraction is easily obtained on the basis of the humidity Y of the gas:

$$y = \frac{Y}{B+Y}$$

where:

$$B = \frac{M_E}{M_G}$$

M_G: mean molar mass of the dry gas: $kg.kmol^{-1}$.

4.2.7. *Differential balances*

We can write the theoretical heat balance in differential form:

$$-G(C_G + YC_V)dt_G = P(C_S + XC_E)dt_P + dE(L_0 + L_L + C_V t_G - C_E t_P) \quad [4.1]$$

The energy of cooling of the gas is abandoned by heat transfer:

$$-G(C_G + YC_V)dt_G = \alpha_S dS(t_G - t_P) \quad [4.2]$$

dS: increment of gas–solid transfer surface.

The mass of solvent evaporated across the surface dS is:

$$dE = M_E C_T \beta_S dS(y_P - y_G) \quad [4.3]$$

In addition:

$$dE = G d Y \quad [4.4]$$

$$d E = - P d X \qquad (dX<0) \qquad\qquad [4.5]$$

Equations [4.2] and [4.3] give us dt_G and dE, and equation [4.1] can be used to calculate dt_P. Equations [4.4] and [4.5] give the variations in humidity of the gas and the product.

EXAMPLE.–

$$t_G = 95°C \qquad G = 0.12 \text{ kg/s} \qquad y_G = 0.019$$
$$t_P = 45°C \qquad P = 0.05 \text{ kg/s} \qquad y_P = 0.032$$
$$dV = 0.1 \text{ m}^3 \qquad \alpha_V = 355 \text{ W.m}^{-3}.\text{K}^{-1} \qquad M_e C_T \beta_V = 0.35$$

$$\left(C_G + YC_V\right) = 1000 \text{ J.kg}^{-1}.°C^{-1}$$

Equation [4.2]:

$$-0.12 \times 1000 \times dt_G = 355 \times 0.1 \times (95 - 45)$$

$$dt_G = -14.8°C$$

Equation [4.3]:

$$dE = 0.35 \times 0.1 \times (0.032 - 0.019) = 0.455 \cdot 10^{-3} \text{ kg.s}^{-1}$$

Equations [4.4] and [4.5]:

$$dY = \frac{0.455 \cdot 10^{-3}}{0.12} = 3.79 \cdot 10^{-3} \qquad dX = \frac{0.455 \cdot 10^{-3}}{0.05} = 0.0091$$

$$dy_G = \frac{3.79 \cdot 10^{-3}}{0.62 + 3.79 \cdot 10^{-3}} = 0.0061$$

y_P can only be calculated using new values of X and t_P.

The number 0.62 is the ratio 18/29 of the molecular masses of water and air.

4.2.8. *Evaporation across an elementary surface*

Consider the following:

W_G: the flowrate of dry gas passing over the surface: $kg.s^{-1}$;

W_V: the flowrate of water vapor passing over the surface: $kg.s^{-1}$;

P: the flowrate of the solid product being renewed along the surface: $kg.s^{-1}$;

ΔS: the elementary surface: m^2.

Let us set:

$$\overline{C}_P = C_P + XC_E \text{ and } \overline{C}_G = C_G + YC_V$$

In addition:

$$y_G = \frac{W_E}{BW_G + W_E} \text{ (with B = 18/29)}$$

$$y_P = \frac{a_E \pi(t_P)}{P_T}$$

a_E: activity of the water at the surface of the solid (it may, in fact, be a solution);

$\pi(t_P)$: saturating vapor pressure of pure water at the temperature t_P: Pa or bar;

P_T: total pressure of the system: Pa or bar.

The heat transmitted by convection warms the product and vaporizes the flowrate of water ΔW_V (first relation):

$$\alpha \Delta S(t_G - t_P) = W_P \overline{C}_P \Delta t_P + L_{PG} \Delta W_V$$

L_{PG}: latent heat of state change (positive) from the liquid at t_P to the vapor at t_G.

The heat transmitted by convection is lost by the gas (second relation)

$$\alpha \Delta S(t_G - t_P) = -(W_G + W_V)\overline{C}_G \Delta t_G$$

α: heat transfer coefficient: $W.m^{-2}.°C^{-1}$.

The evaporation of water across the surface ΔS corresponds to the flowrate ΔW_E.

$$\Delta W_E = c_T \beta \Delta S(y_P - y_G)$$

β: material transfer coefficient: $m.s^{-1}$.

In view of the ideal gas law, the concentration of a gas is:

$$c_T = \frac{n}{V} = \frac{P_T}{RT} \qquad (kmol.m^{-3})$$

Hence (in $kg.m^{-2}.s^{-1}$), we obtain the third relation:

$$\Delta W_E = \frac{\beta M_E \Delta S P_T}{RT}\left(\frac{a_E \pi(t_P)}{P_T} - \frac{W_E}{W_E + BW_G}\right)$$

M_E: molar mass of water: 18 $kg.kmol^{-1}$.

Thus, we have 3 relations to find the evolution of the following three variables.

t_P, t_G and W_E

This system can only be solved numerically for a dryer, working slice by slice.

4.2.9. Wet temperature

We shall define the wet temperature for any given gas passing over the surface of any solution.

The differential energy balance is written:

$$-G(C_G + YC_V)dt_G = P(C_S + XC_E)dt_P + LdE$$

C_S: specific heat capacity of the solute: $J.kg^{-1}.°C^{-1}$.

In this expression:

$$L = L_0 + C_V t_G - C_E t_P$$

L_0: latent heat of vaporization at 0°C: $J.kg^{-1}$.

From the formulation of the balance, we draw:

$$dt_P = \frac{-LdE - G\overline{C}_G dt_G}{P(C_S + XC_E)} \qquad [4.6]$$

where:

$$\overline{C}_G = C_G + YC_V$$

However, as the product heats up, t_P increases, but:

$$dE = M_E c_T \beta_S (y_S + y_G)dS$$

If t_P increases, y_P also increases and dE does too. However, according to equation [4.6], dt_P decreases and, finally, *t_P tends toward a limit, which is the wet temperature t_{PH}* that we are seeking. The differential balance becomes:

$$-G(C_G + YC_V)dt_G = LdE$$

However, the cooling of the gas is due to heat convection:

$$-G(C_G + YC_V)dt_G = \alpha_S(t_G - t_{PH})dS$$

In addition, the transfer of latent heat is written:

$$LdE = LM_E C_T \beta_S (y_{PH} - y_G)dS$$

Thus, the differential heat balance becomes:

$$\alpha_S \left(t_G - t_{PH} \right) = L M_E C_T \beta_S \left(y_{PH} - y_G \right)$$

Remember the meaning of the various symbols.

L: heat of vaporization of water: $J.kg^{-1}$;

M_E: molar mass of water: $kg.kmol^{-1}$;

c_T: total molar concentration of the gaseous phase: $kmol.m^{-3}$

$$c_T = \frac{P_T}{RT}$$

P_T: total pressure: Pa;

R: ideal gas constant: $8314\ J.kmol^{-1}.K^{-1}$;

T: absolute temperature: K;

β_S: surface material transfer coefficient: $m.s^{-1}$

$$\beta_S = \frac{\alpha_S \sigma}{C_G \rho_G}$$

α: surface heat transfer coefficient: $W.m^{-2}.K^{-1}$;

σ: psychrometric coefficient: see section 4.3.3;

C_G: specific heat capacity of the gaseous phase: $J.kg^{-1}.°C^{-1}$;

ρ_G: density of the gaseous phase: $kg.m^{-3}$.

Thus, the differential heat balance is written, after simplification by α_S and if we accept the ideal gas law:

$$t_G - t_H = \frac{L M_E c_T \sigma}{C_G \rho_G} \frac{\left(a_E \pi(t_H) - P_V \right)}{P_T}$$

For water vapor and air:

$$L \# 2.4.10^6 \, \text{J.kg}^{-1}$$

$$C_G \# 1000 \, \text{J.kg}^{-1}.^\circ C^{-1}$$

$$\sigma \# 1$$

4.2.10. *Calculation for a continuous-flow dryer in the convection regime*

To perform this calculation, we need to adopt the following approach:

1) find the overall theoretical heat balance of the dryer;

2) starting at the inlet of the product (supposed to circulate in a countercurrent against the air), perform the calculations slice by slice;

3) we cease the calculation when the temperature of the gas is equal to its temperature on input into the device.

In establishing the overall heat balance, we must choose the temperatures at the ends of the device so that the gap between the temperature of the gas and that of the product is never any less than 10°C.

4.2.11. *Drying of crystals and grains (a possible solution)*

When we dry divided solids (crystals, in particular), it could be supposed that the transfer surface remains constant and the temperature of the liquid surface is equal to the humid temperature of the air. If the temperature of the air does not vary, the evaporation rate should remain constant as well. However, this is not actually the case, because at the surface of each crystal, there are wet zones which are in contact with dry zones and, in reality, the evaporation surface constantly decreases. We can agree, finally, that the thickness of humidity remains constant and equal to e_E and that, consequently, the evaporation surface is proportional to the humidity.

Let A_F represent the initial surface area of a sample of product to be processed, whose mass is M_{PF}, and let X_F be the humidity of that product (ρ_L is the density of the liquid). The thickness of the wet layer e_E is then:

$$e_E = \frac{X_F M_{PF}}{\rho_L A_F}$$

In a slice of a dryer, the mass of product present is $P \, \Delta\tau$ if $\Delta\tau$ is the time taken to traverse the slice. The variation in surface area in that slice, then, is:

$$\Delta A = \frac{\Delta X P \Delta\tau}{\rho_L e_E}$$

The drying rate then decreases exponentially, as is proved by the drying, in a fluidized bed, of a load composed of sucrose crystals (see [MOR 95]). Similar tests conducted on grains of corn showed the same result. It is undeniably possible to interpret this result by saying that the transfer coefficient decreases in a linear fashion as the humidity does. This is an illusory but simple method of expressing the migration of the liquid from the interior of the product to its surface.

The experiments run by Mourad *et al.* [MOU 95] confirmed this exponential decrease.

4.2.12. *A possible calculation method for continuous-flow dryers*

If $\overline{V}_p$ is the mean rate of progression of the product through the dryer, certain authors have introduced what is called the "slowness" $\lambda = 1 / V_p$, whose mean value is obviously:

$$\overline{\lambda} = 1 / \overline{V}_p$$

Suppose, then, that the slowness (or speed) obeys a distribution law whose frequency $f(\lambda)$ may obey a Gauss normal law, a Poisson law, a normal logarithmic law or any other law. The law is supposed to be normalized:

$$\int_{0}^{\infty} f(\lambda) d\lambda = 1$$

Let us choose a number n between 6 and 8 and set:

$$\Delta\lambda = \overline{\lambda} / n$$

For the variation of λ, we consider n intervals of breadth $\Delta\lambda$. Let λ_i represent the value of λ in the middle of the interval with index i.

Now suppose that the dryer is fed respectively at the flowrates P and G for the product and gas. Let us divide the dryer into n partial dryers, working in parallel. The partial dryer with index i will be fed by the flowrates:

$$Pf(\lambda_i)\Delta\lambda \quad \text{and} \quad Gf(\lambda_i)\Delta\lambda$$

Each partial dryer can be calculated independently of the others, and for this, the output temperature is t_{Pi}, whilst the residual humidity is X_i. The overall conditions at the output of the dryer, then, are:

$$X_{output} = \sum_{1}^{n} f(\lambda) X_i \Delta\lambda \quad \text{and} \quad t_{output} = \sum_{1}^{n} (\lambda) t_{pi} \Delta\lambda$$

4.2.13. *Irreducible humidity*

As the product advances in the dryer, its humidity decreases. When that humidity becomes such that it corresponds to a thickness of parent liquor equal only to 150 molecular diameters of solvent, the nature of the phenomenon changes, and we are then dealing with a surface adsorption which does not obey the same laws.

We shall now use an example to see which humidity that criterion corresponds to.

Let us assimilate the crystals to spheres 200 μm in diameter. Remember that *the diameter of the water molecules is around 0.35 nm*. Suppose that the density of the crystals is equal to 2000 kg/m³ and that of the parent liquor to 1000 kg/m³.

The mass of a crystal is:

$$\frac{\pi}{6}\left(2\times10^{-4}\right)^{3}\times=2000=8.4\times10^{-9}\ \text{kg}$$

The surface of a crystal is:

$$\pi\left(2.10^{-4}\right)^{2}=1.26.10^{-7}\ \text{m}^{2}$$

The mass of liquor present is:

$$1.26.10^{-7}\times150\times0.35.10^{-9}\times1000=6.6.10^{-12}\ \text{kg}$$

The corresponding humidity is:

$$6.6.10^{-12}/8.4.10^{-9}=9.10^{-4}\ \#1^{\circ}/\circ\circ$$

Of course, this irreducible humidity increases with the fineness of the particles. It is of the order of magnitude of the guarantees offered by the dryer manufacturers. If the crystals have water of hydration (or solvation), the residual humidity may surpass this irreducible humidity.

When the irreducible humidity is reached, the phenomenon is considerably simplified. It is then a case of a simple transfer of heat between two phases without evaporation.

4.2.14. *Enthalpy graph*

It is possible to summarize the previous considerations with an enthalpy graph, such as that shown in Figure 4.3.

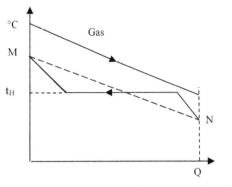

Figure 4.3. *Enthalpy graph*

On the ordinate axes, we place the temperatures of the gas and product, and on the abscissa axis the thermal power transferred Q. The platform characterizing the evolution of t_p corresponds to the humid temperature of the system.

4.2.15. *Quick and approximate estimation of a dryer*

Experience in the industrial sector shows that, for an approximate estimation, we can assimilate a dryer to a purely thermal exchanger working with two products (a gas and a solid) in direct contact, for cocurrent or countercurrent. On the enthalpy graph in Figure 4.3, this is equivalent to linking the points M and N by a straight line.

The thermal power transferred is then simply:

$$Q = \alpha_x \Omega \Delta T$$

α_X: heat transfer coefficient expressed in relation to the value Ω, which may be the surface area of the solid particles or the volume of the dryer (α_S is measured in $W.m^{-2}.°C^{-1}$ and α_V in $W.m^{-3}.°C^{-1}$);

Ω: surface or volume;

ΔT: LMTD (logarithmic mean temperature difference) between the gas and the product to be dried.

In view of the theoretical balance:

$$Q = G\left(C_G + \bar{Y}C_V\right)\left(t_{G0} - t_{G1}\right)$$

where:

$$\bar{Y} = \frac{1}{2}\left(Y_0 + Y_1\right)$$

Similarly:

$$Q = P\left(C_P + \bar{X}\, C_E\right)\left(t_{P1} - t_{P0}\right) + \Delta E\left(L_0 + C_V t_{G1} - C_E t_{P0} + \bar{L}_L\right)$$

$\bar{X}$: mean humidity of the product between inlet and outlet.

This risky method is called the "equivalent exchanger method".

4.2.16. *Diffusional drying*

To verify whether a drying operation takes place in the diffusion regime, we use the following criterion:

$$Bi > 100$$

The Biot number is defined by:

$$Bi = \frac{\beta d}{D}$$

The diffusivity D is of the order of 10^{-9} m^2/s. The parameter d here is the diameter of a particle supposed to be spherical.

The material transfer coefficient β is obtained on the basis of the heat transfer coefficient α. For simple currents of air lapping at a surface, α is of the order of 10 W.m^{-2}.°C^{-1}.

Hence, for water vapor in air:

$$\beta = \frac{\alpha\sigma}{C_G\rho_G} = \frac{10 \times 1}{1000 \times 1.29} = 7.75 \times 10^{-3} \text{ m.s}^{-1}$$

and:

$$Bi = \frac{7.75 \times 10^{-3} \times 1.5 \times 10^{-3}}{10^{-9}} = 1.16 \times 10^{4} \gg 100$$

(We have chosen a diameter grain equal to 1.5 mm).

Thus, the hypothesis is borne out, and we can say that the surface of the solid is at humidity equilibrium with the ambient gas.

However, we can state that the temperature is uniform within the particles of product, because the ratio between the heat diffusivity and material diffusivity is:

$$\frac{\lambda}{C\rho D} -- = \frac{0.2}{2000 \times 1000 \times 10^{-9}} \#100$$

The heat diffusivity is 100 times greater than the material diffusivity. In addition, the drying of foods and plants takes place in a gaseous atmosphere at a very moderate temperature so as not to damage the product.

The diffusion equations are identical whether we are dealing with heat or material.

The building blocks for the solutions to these equations are to be found in [LON 85].

4.3. Transfer coefficients

4.3.1. Heat transfer coefficient

1) Along a flat surface

If the air shifts parallel to the surface of the solid:

$$\alpha = 14,4\,G^{0,8}$$

α: $W.m^{-2}.°C^{-1}$;

G: $kg.m^{-2}.s^{-1}$ with: $G = \rho_G V_G$;

ρ_G: air density: $kg.m^{-3}$;

V_G: air speed: $m.s^{-1}$.

2) Around a sphere

We calculate the Nusselt number:

$$Nu = \frac{\alpha d}{\lambda_G} = 2 + 0,6\,Re^{0,5}\,Pr^{0,33}$$

Re is the Reynolds number:

$$Re = \frac{V_G d\rho_G}{\mu_G}$$

d: diameter of the sphere: m;

μ_G: viscosity of the gas: Pa.s.

[FRÖ 38] performed a detailed study of the evaporation of a liquid drop in a stream of gas.

3) Across a fixed bed

In line with [WIL 45], we calculate:

$$Re = \frac{V_G d\rho_G}{\mu_G}$$

d: particle diameter: m;

V_G: velocity of the gas in an empty bed: $m.s^{-1}$.

We define:

$$j_H = \frac{\alpha}{C_G \rho_G V_G}$$

C_G: gravimetric specific heat capacity of the gas: $J.kg^{-1}.°C^{-1}$

$$Re < 350 \qquad\qquad j_H = 1.82\ Re^{-0.51}$$
$$Re > 350 \qquad\qquad j_H = 0.989\ Re^{-0.41}$$

4) Across a fluidized bed

According to Kunii and Levenspiel:

$$\alpha = 0.03\frac{\lambda_G}{d_P}Re^{1/3}$$

d_P: particle diameter: m;

λ_G: heat conductance of the gas: $W.m^{-1}.°C^{-1}$.

$Re = \dfrac{\rho_G V_G d_P}{\mu_G}$ Re is the Reynolds number.

5) Jets perpendicular to a surface

According to [KER 68]:

– Circular orifices of diameter d:

If e is the distance from the orifice to the surface:

$$e/d \leq 5 \qquad\qquad Nu = 0,075\ Re^{0,745}$$
$$e/d > 5 \qquad\qquad Nu = 0,32\,Re^{0,745}\,(e/d)^{-0,828}$$

where:

$$Nu = \frac{\alpha d}{\lambda_G} \qquad\qquad Re = \frac{V_G d\rho_G}{\mu_G}$$

V_G: velocity through the orifice: $m.s^{-1}$.

– Rectangular gaps:

l: width of the gaps: m;

e: distance from the gaps to the surface: m

$e/l \leq 5$ $Nu = 0.901 \, Re^{0.437}$

$e/l > 5$ $Nu = 0.135 \, Re^{0.697} \, (e/l)^{-0.351}$

The definitions of Nu and Re are the same as with the orifice, but where $d = 2\, l$.

4.3.2. *Correspondence between surface and volume coefficients*

The coefficients α and β, which we have just defined, express the transfers of heat and material across the gas–solid interface.

However, to run the calculations for certain devices, it is helpful to have coefficients expressed in relation to the volume of suspension of the solid in the gas. The relation linking a surface coefficient α_S or β_S to the corresponding volume coefficient α_V or β_V is simple.

$$\alpha_V = a\,\alpha_S \quad \text{and} \quad \beta_V = a\,\beta_S$$

a is the surface area of solid per unit volume of suspension. Thus:

1) fluidized bed:

$$a = \frac{6(1-\varepsilon)}{d_P}$$

d_p: diameter of the fluidized particles: m;

ε: empty fraction (porosity) of the bed (which increases with the velocity of the gas);

2) pneumatic dryer:

We also have:

$$a = \frac{6(1-\varepsilon)}{d_P}$$

$(1-\varepsilon)$ results from the knowledge of the respective flowrates G and P of gas and product (in $kg.s^{-1}$):

$$(1-\varepsilon) = \frac{P/\rho_P}{\dfrac{P}{\rho_P} + \dfrac{G}{\rho_G}}$$

Indeed, in a device such as this, the velocities of the gas and the product are similar.

ρ_G and ρ_P are the densities $(kg.m^{-3})$ of the gas and the solid product;

3) rotating drum:

We characterize the apparent volume of solid present in the drum by the degree of filling R, which is the ratio of that volume to the drum's total internal volume. Thus, we have:

$$a = \frac{6R(1-\varepsilon)}{d_P}$$

4.3.3. *Material transfer coefficients. Psychometric coefficient*

The Chilton–Colburn analogy means we can write:

$$\beta Sc^{2/3} = \frac{\alpha}{C_G\rho_G} Pr^{2/3}$$

Thus:

$$\beta = \frac{\alpha}{C_G\rho_G}\left[\frac{Pr}{Sc}\right]^{2/3} = \frac{\alpha}{C_G\rho_G}(Lu)^{2/3} \qquad (m.s^{-1})$$

Lu: Luikov number (also known as the Lewis number):

$$Lu = \frac{DC_G\rho_G}{\lambda_G}$$

D: diffusivity of the solvent in the gas: $m^2.s^{-1}$;

λ_G: heat conductivity of the gas: $W.m^{-2}.°C^{-1}$;

C_G: gravimetric specific heat capacity of the gas: $J.kg^{-1}.°C^{-1}$.

We can set:

$$Lu^{2/3} = \sigma$$

σ is the psychrometric coefficient.

For the air–water system, σ ranges from 1.1 (if $Y_G = 0.01$) to 1.05 (if $Y_G = 0.1$).

The Luikov number, for that system, is equal to 1.17.

NOTE.–

Consider the inside of a tube with the diameter D and length L. In the turbulent regime, [COL 31] write:

$$\alpha = \frac{1}{2} f \rho \, V \, C \qquad \text{and} \qquad \Delta P = 2f \rho \, V^2 \frac{L}{D} = 4f \frac{\rho V^2}{2} \frac{L}{D}$$

C: specific heat capacity of the fluid: $J.kg^{-1}.°C^{-1}$;

V: flow velocity of the fluid in an empty bed: $m.s^{-1}$;

ρ: density of the fluid: $kg.m^{-3}$.

The dimensionless friction coefficient f depends only on the Reynolds number.

4.4. Capillary pressure (imbibition/drainage)

4.4.1. Definition

Consider two immiscible fluids M and m which, between the two of them, completely fill the voids in a porous solid. The fluid M wets the solid more fully than the fluid m.

At all points in the solid, the two fluids are separated by an interface whose main curvatures are $\dfrac{1}{R}$ and $\dfrac{1}{r}$, where $|R| > |r|$.

We shall agree that:

– if the center of a curvature is situated in the fluid m which wets the solid less well than the fluid M, that curvature will be counted positively;

– the capillary pressure is defined by:

$$P_c = P_m - P_M$$

Laplace (see [POI 95]) showed that the capillary pressure is written:

$$P_c = \gamma\left(\frac{1}{R} + \frac{1}{r}\right)$$

Let us now look at two examples to show that these two conventions are coherent.

When the water rises in a capillary tube with a circular cross-section and the water wets the wall of the tube, the capillary pressure is (here $R = r$)

$$P_c = \frac{2\gamma}{R} > 0$$

That is:

$$P_m > P_M \text{ or indeed } P_{air} > P_{water}$$

Now consider the liquid ring which joins two solid particles in contact with one another.

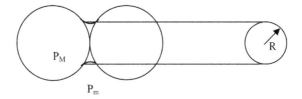

Figure 4.4. *Liquid ring joining two particles*

Using this convention, the capillary pressure can be written as:

$$P_c = \gamma\left(\frac{1}{R} + \frac{1}{r}\right)$$

The surface of the liquid ring in Figure 4.4 comprises:

– a short radius of curvature, r, which, accordingly, is counted positively;

– a long radius of curvature, R, which is counted negatively. We therefore have:

$$P_c = \gamma\left(\frac{1}{|r|} - \frac{1}{|R|}\right) > 0 \quad \text{and} \quad P_M < P_m$$

The liquid ring is subject to an external capillary pressure which maintains its cohesion, i.e. the contact between the two particles. Indeed, on the surface of the particles wetted by the liquid M, the pressure P_M is low, which tends to draw the particles closer together.

[MEL 67] and [PIE 67] reported the geometric characteristics and adhesion force of the meniscus between two spheres.

4.4.2. Contact angle

Consider the triple point at which the two fluids and the solid S meet.

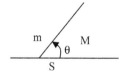

Figure 4.5. *Contact angle*

In this case, the fluid M will wet the solid more than the fluid m because the angle θ is less than π/2.

4.4.3. *The three states of a porous material imbided with a liquid*

We shall define that "saturation" as the ratio of the volume of liquid to the total volume of voids in the porous material. A distinction is drawn between:

1) The saturated state. Saturation is equal to 1.

2) The pendular state. This is the state with the lowest level of saturation. The liquid takes the form of rings binding the solid particles to one another. We can consider that in each pair of particles, one of the particles is suspended from the other by a liquid ring, much like a pendulum (see Figure 4.6).

3) The funicular state, from the Latin word "*funiculus*" (chord). The addition of water in the pendular state causes the liquid rings to increase in size and coalesce. The result is a continuous liquid structure which more or less fills the voids in the porous substance. As such, it is possible to move *through the liquid* from one point to another via a tortuous route that could be likened to an intertwined piece of string.

In the pendular state, the most common menisci are shown in Figure 4.6.

Figure 4.6(a) shows the liquid ring linking two spheroidal particles. The curvatures are in opposite directions and their respective sizes are such that the pressure within the liquid is less than the external pressure on the liquid, which causes the two particles to bind together. If the liquid evaporates, the two radii of curvature tend towards zero, so the difference ΔP increases, but the surface of contact between the liquid and the solid decreases.

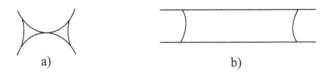

a)   b)

Figure 4.6. *Arrangement of the menisci in the pendular state*

Figure 4.6(b) shows a pore that is open at both ends with the cylindrical volume of liquid filling it, at least partially. When the liquid evaporates, the shape of the menisci does not change.

4.4.4. *Capillary pressure and saturation*

The mercury impregnation method is used to determine the function P_C (S).

The porous material, which is totally saturated with water (S = 1) at ambient pressure, is immersed in mercury. The pressure of the mercury is gradually increased, whilst we measure the change in mass of the porous material. Meanwhile its volume, which is known, remains constant. The density of the material thus impregnated with water and mercury is then:

$$\frac{M}{V} = (1-\varepsilon)\rho_S + S\varepsilon\rho_L + (1-S)\varepsilon\rho_{Hg}$$

In this relation, ρ_S, ρ_L and ρ_{Hg} are the known densities of the solid, water and mercury respectively. Consequently, from this same formula, we can determine the saturation S corresponding to the pressure of the mercury, which is, in fact, equal to the capillary pressure. From these measurements, we can derive the function P_C (S), the shape of which is illustrated in Figure 4.7.

A satisfactory device should be capable of achieving P_C values of the order of a few hundred bars.

In practice, P_C (S) is of the form:

$$P_C = \frac{A}{\left(S - S_I\right)^{1/\lambda}}$$

S_I is the irreducible saturation.

By extending the curve (in Figure 4.7) to the abscissa point where S = 1, we can obtain the input capillary pressure P_E that must be overcome before the liquid begins to penetrate a totally saturated body.

Due to the existence of partially-open pores, the irreducible saturation is specific to the impregnation method. Indeed, the operation of pre-drying leaves some of the partially open pores full, which, thermodynamically, they should not be.

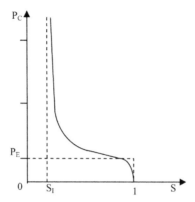

Figure 4.7. *Change in P$_C$ as a function of saturation (imbibition)*

However, the influence of the capillary phenomena is only notable at high saturation levels. Consequently, the irreducible saturation can be discounted, and the formula can simply be written as:

$$P_C = \frac{P_P}{S^{1/\lambda}}$$

In this expression, P$_P$ is the fracture capillary pressure. To determine λ, we merely need to considerer saturations greater than 0.4.

4.5. Thermodynamics and capillarity

4.5.1. *Free energy and interfacial area*

Where capillarity is concerned, it is customary to relate free energy to the interfacial area. We shall adopt a different approach.

Suppose that the pressure of the non-wetting fluid is the same within and outside the porous substance, and consider an elementary volume dV of wetting fluid transferred from the porous substance to a flat surface. Its pressure will increase by P$_C$.

$$\frac{dF}{dV} = \left[-\left(-P_C\right) + 0 \right] = P_C$$

The wetting-liquid saturation will have changed by $-dS_M$.

The Helmholz energy of the volume $dV = dS_M$ (where $dS < o$) will have changed by:

$$dF = P_C \ dV = -P_C \ dS \quad > 0$$

However, we know that:

$$dF = \gamma d\sigma$$

where σ is the interfacial area.

The variation in this area is then:

$$\sigma_2 - \sigma_1 = \frac{-1}{\gamma} \int_{S_{M1}}^{S_{M2}} P_C dS_M$$

In practice, it is possible that the pressure of the less wetting fluid m may vary depending on whether we are dealing with imbibition or drainage, and this will give different values for P_C (see section 4.7).

4.5.2. *Kelvin equation*

Consider the flat surface of a liquid above which the gaseous phase is at pressure P_0 and *comprises only the vapor of the liquid*.

If a vertical tube is plunged into the liquid, two scenarios are possible:

– the liquid rises in the tube above the initial level (Figure 4.8(a));

– the tube remains partially empty below the initial level (Figure 4.8(b)).

a) Raised meniscus: the surface tension draws the liquid column upward and this force is compensated by the weight of the column.

b) Lowered meniscus: the surface tension draws the liquid column downward and this force is compensated by the upward force of buoyancy (Archimedes' principle).

In both cases:

$$2\pi r\gamma\cos\theta = P_C hg\pi r^2$$

The difference in pressure between the meniscus and the initial plane is:

$$\Delta P = P_1 - P_0 = -\rho_L gh = \frac{-2\gamma\cos\theta}{r}$$

r: radius of the tube: m;

P_1: pressure at the meniscus: Pa;

P_0: saturating vapor pressure on a flat surface: Pa;

γ: surface tension: $N.m^{-1}$;

θ: contact angle: rad;

ρ_L: density of the liquid: $kg.m^{-3}$;

g: acceleration due to gravity: $9.81 \ m.s^{-2}$.

Indeed, in scenario A, taking the initial plane as a starting point, the pressure of the liquid will decrease as we move up towards the meniscus and take the value P_1.

In scenario B, the hydrostatic pressure will increase up to the meniscus and take the value P_1.

Now consider an ideal gas. Its chemical potential is:

$$\mu_G = \mu_0 + RT \ Ln \ P$$

Suppose that this gas is at thermodynamic equilibrium with the liquid (the gas is the vapor of the liquid).

Along the height h, the change in chemical potential of the liquid is:

$$\Delta\mu_L = V_L\Delta P = \frac{-M}{\rho_L} \times \frac{2\gamma\cos\theta}{r} \qquad \text{(Laplace law)}$$

The variation in chemical potential of the vapor is:

$$\Delta\mu_G = RTLn\frac{P_1}{P_0} \qquad \left(\Delta\mu_G = \int_{P_0}^{P_1} V_c dP \right)$$

These two variations are equal. Accordingly:

$$P_1 = P_0 \exp\left(-\frac{M_L}{\rho_L}\frac{2\gamma\cos\theta}{RTr}\right)$$

This latter relationship is known as the Kelvin equation, which was formulated approximately by [THO 71], but not demonstrated.

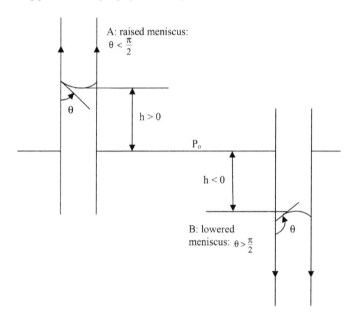

Figure 4.8. *Position of the meniscus with respect to the initial plane*

4.5.3. Consequence: liquid film on a solid surface

The vapor pressure P of the liquid will be:

$$Ln\frac{P_1}{P_0} = \frac{M\gamma}{RT}\left(\frac{1}{R}+\frac{1}{r}\right)$$

Let us look at four specific cases.

1) Value of P_1 at the inner surface of a cylinder. P_1 is on the concave side. Thus:

$$P_1 < P_0 \text{ and } Ln\frac{P_1}{P_0} = -\frac{M\gamma}{RT}\left(\frac{1}{r}\right) < 0$$

r: radius of the cylinder: m.

Indeed, $1/R = 0$.

[COH 38] directly justified the application of the Kelvin equation to this case.

2) Full cylinder. Value of P_1 at the outer surface of the cylinder, on the convex side.

$$P_1 > P_0 \quad \text{and} \quad Ln\frac{P_1}{P_0} = \frac{M\gamma}{RT}\left(\frac{1}{r}\right) > 0$$

3) Full sphere of radius r ($R = r$). The vapor is on the convex side:

$$P_1 > P_0 \quad \text{and} \quad Ln\frac{P_1}{P_0} = \frac{2M\gamma}{RT}\left(\frac{1}{r}\right) > 0$$

4) Hemispherical tip of a solid shaft. The radius of the tip is r:

$$P_1 > P_0 \quad \text{and} \quad Ln\frac{P_1}{P_0} = \frac{2M\gamma}{RT}\left(\frac{1}{r}\right) > 0$$

NOTE.–

As long as adsorption occurs on the walls of the pores (considered to be cylindrical), the difference between the partial pressure P_V of the adsorbate and the saturating vapor pressure is:

$$P_V - P_0 \exp\left(\frac{-M\gamma}{RTr}\right)$$

When, due to narrowing in the pores, menisci form suddenly, the pressure difference becomes:

$$P_V - P_0 \exp\left(\frac{-2M\gamma}{RTr}\right)$$

Condensation then occurs much more quickly.

4.5.4. Equilibrium relation (the gas contains an inert substance)

According to the ideal gas hypothesis, the chemical potential of a vapor in a gas at molar fraction y is:

$$\mu_V\left(T, P_G, y\right) = \mu_V^*\left(T, 1\right) + RTLn\left(P_G y\right)$$

If P_G is measured in bar, μ_V^* is the chemical potential of the vapor under the following conditions:

– $P_G y = 1$ bar;

– T degrees Kelvin.

At the same temperature T, but at the pressure P_L, the chemical potential of a pure, incompressible liquid is:

$$\mu_L\left(P_L, T\right) = \mu_L^*\left(T, 1\right) + \frac{M}{\rho_L}\left(P_L - 1\right) \qquad \left(J.kmol^{-1}\right)$$

M and ρ_L : molar mass and density of the liquid.

If P_L is measured in bar, μ_L^* is the chemical potential of the pure liquid under the following conditions:

– $P_L = 1$ bar;

– T degrees Kelvin.

If the liquid is separated from the vapor by a flat surface, we know that, at equilibrium, the partial vapor pressure P_{GY} is equal to the saturating vapor pressure of the liquid $\pi(t)$. Moreover:

$$P_L = \pi(T)$$

The equality of the chemical potentials is written as:

$$\mu_V^* (T,1) + RTLn\pi = \mu_L^* (T,1) + \frac{M}{\rho_L}(\pi - 1) \qquad [4.7]$$

This equation defines a relation between π and T:

$$\pi = f(T)$$

At equilibrium, if $P_G \neq P_L$, the equality of the chemical potentials is written as:

$$\mu_V = \mu_L$$

That is:

$$\mu_V^* (T,1) + RT Ln P_G y = \mu_L^* (T,1) + \frac{M}{\rho_L}(P_L - 1) \qquad [4.8]$$

Subtracting equation [4.7] from equation [4.8], term by term, we obtain:

$$RT Ln \frac{P_G y}{\pi} = \frac{M}{\rho_L}(P_L - \pi)$$

If, on the other hand, the two phases are separated by a curved surface, according to Laplace's law their respective pressures would be different and, if, as is generally the case, the concavity of the surface faces toward the gaseous phase, then:

$$P_L = P_G - P_C$$

The pressure P_C, which is positive, is the capillary pressure.

Hence:

$$P_G y = \pi \exp\left[\frac{-M}{RT\rho_L}\left(P_C + \pi - P_G\right)\right]$$

$(\pi - P_G)$ is non-zero if *the gaseous phase contains an inert gas in addition to the saturating vapor.*

This equation represents the local equilibrium relation.

In practice, drying is never performed under pressure, but rather at atmospheric pressure or even in a vacuum. Thus, the pressure P_G does not exceed 1 bar abs.

The temperature of the product, and thus of the gaseous mixture within its pores, varies between the ambient temperature and a maximum of 150°C. Consequently, for water, $\pi(T)$ is greater than 1 or 3 kPa and less than 5 bar abs.

We shall see that the orders of magnitude of the capillary pressure P_C, for pore radii of between 10 and 30 µm, fall between 140 bar and 4800 Pa (see Table 4.2 in section 4.7.2).

Thus, we can see that it is prudent not to discount the term $(\pi - P_G)$ in the exponential function when P_C is low – i.e., for grain sizes greater than or equal to 1 µm.

At 140°C:

$$\frac{RT\rho_L}{M} = \frac{8314 \times 413 \times 1000}{18} = 1.9.10^8 \, Pa$$

If $P_C = 140$ bar, the result of the exponential function is 0.93, which is not far from 1. However, this does not imply that this correction is negligible.

The relative humidity can be derived as follows:

$$\psi = \frac{P_G y_v}{\pi} = \exp\left[\frac{M}{RT\rho_L}\left(P_G - \pi - P_C\right)\right]$$

If we know the temperature T, the saturation S_C – and thus the capillary pressure and the total pressure P_G of the gaseous phase – it is easy to calculate ψ corresponding to a given level of saturation.

Conversely, if the values of ψ, T (thus π) and P_G are known, we can calculate P_C, and thus the capillary saturation S_C (see section 4.4.4).

Bear in mind that the relative humidity of the gaseous phase at equilibrium with a liquid surface that is concave on the side of the gas diminishes as P_C increases, i.e. as the radius of curvature of the surface decreases.

Capillary condensation therefore occurs firstly on the walls of those pores with the smallest diameters as these pores have been wetted beforehand by simple adsorption.

4.6. Adsorption and capillarity

4.6.1. *Adsorption and capillarity in an inorganic porous substance*

The combined effects of adsorption and capillarity that we have just analyzed in light of the work of El-Sabaawi *et al.* [EL 77] are illustrated in Figure 4.9.

In the case of humidification, and on account of the phenomenon of hysteresis, the arc of the curve representing capillarity alone corresponds to lower values of saturation S and is not shown.

El Sabaawi *et al.* [EL 77] experimentally verified the considerations above on catalysts.

Pure adsorption isotherms can be represented using Rounsley's equation [ROU 61]:

$$\Omega_e = \frac{MN_0 C\psi}{\left[1+(C-1)\psi\right]}\left[\frac{1-\psi^n}{1-\psi}\right] = \frac{BC\psi}{1+(C-1)\psi}\left[\frac{1-\psi^n}{1-\psi}\right]$$

where:

$$C = \exp\left[\frac{E - \Lambda}{RT}\right]$$

Ω_e: volume of water adsorbed per m² of porous solid;

E: molar heat of adsorption of the first layer: J.kmol⁻¹;

Λ: heat of condensation: J.kmol⁻¹;

ψ: relative humidity: P_{GY}/π;

N_0: number of kmoles adsorbed on 1m² in a mono-molecular layer: m⁻²;

M: molar mass of water: kg.kmol⁻¹;

n: number of adsorbed layers;

ρ_L: density of water in the liquid state: kg.m⁻³.

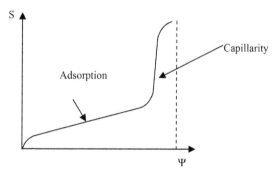

Figure 4.9. Function ψ (S) for the drying
of an inorganic porous substance

Let ε be the total porosity (empty fraction) of a dry, homogeneous, porous solid. The liquid saturation S is the fraction of the porosity occupied the liquid. Thus, per m³ of porous solid, the volume of the liquid is $S\varepsilon$.

The saturation S of the porous body is the sum of two terms:

– capillary saturation S_C;

– adsorption saturation S_A.

As we have seen, if ψ is known, we can calculate P_C which, itself, gives the capillary saturation S_C (see section 4.4.4).

Since the saturation S_C varies between 0 and 1, dividing this interval into n equal parts gives us:

$$\Delta S = I / n$$

$$S_{C,i} = i\Delta S$$

The values $S_{C, i-1}$ and $S_{C,i}$ correspond to the values $P_{C,i-1}$ and $P_{C,i}$.

The index i varies between 1 and N, increasing as S increases.

Let us set:

$$\overline{P}_{C,i} = \frac{P_{C,i-1} + P_{C,i}}{2}$$

Laplace's law gives us:

$$\overline{R}_i = \frac{2\gamma}{\overline{P}_{C,i}}$$

The total equivalent length L_i of the pores of radius $\overline{R}_i$ is:

$$L_i = \frac{\Delta S}{\pi \overline{R}_i^2}$$

The corresponding surface is:

$$a_i = 2\pi \overline{R}_i L_i = \frac{2\Delta S}{\overline{R}_i} = \frac{\Delta S \overline{P}_{C,i}}{\gamma}$$

The surface A_i available for adsorption is:

$$A_i = \sum_{j=i+1}^{n} a_j$$

Consequently, the total saturation S is:

$$S = S_{C,i} + A_i \Omega_e (\psi, T) = S_{C,i} + S_{A,i}$$

Ω_e: volume of adsorbed water per m^2 of porous solid.

If S is high, then $S_{C,i}$ is also high, which means that i is close to n. In this case, A_i is low, as is $S_{A,i}$.

Thus, if we consider a pressure P_G, a temperature T, and a relative humidity ψ, it is possible to calculate the total saturation S.

4.7. Hysteresis

4.7.1. *Hysteresis. Adsorption and desorption of a gas in an inorganic solid*

Consider the following relative humidity:

$$\varepsilon = P_V / \pi(t)$$

P_V: partial vapor pressure of water in the gas;

π: saturating vapor pressure of water.

Above a critical level of relative humidity ε_{cr} in the gaseous phase, the sorption exhibits hysteresis due to capillary phenomena within the pores, as illustrated by Figure 4.10.

In the following industrial operations, *the relative humidity ε remains less than ε_{cr}*:

– preparative chromatography (seldom used for gases);

– purification of gaseous mixtures by adsorption of a compound to a particulate solid.

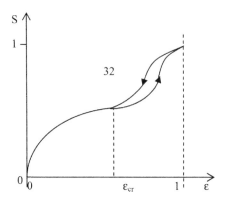

Figure 4.10. *Adsorption-desorption hysteresis*

4.7.2. *Curvature and pore radius. Hysteresis by contact angle*

The contact angle θ is the angle of the interface with the wall of a pore supposed to be cylindrical. The relationship between the radius of curvature R_I of the interface and the radius of the tube R_T is established by the equation:

$$R_T = R_I \cos \theta$$

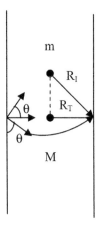

Figure 4.11. *The interface in a pore*

The capillary pressure is:

$$P_C = \frac{2\gamma}{R_I} = \frac{2\gamma\cos\theta}{R_T} = P_m - P_M$$

Table 4.2 shows a few values of P_C for water with a surface tension of 0.072 kg.s^{-2}.

R_I	10 nm	0.1 μm	1 μm	30 μm
P_C	140 bar	14 bar	1.4 bar	4800 Pa

Table 4.2. *Capillary pressure*

Note that R_I depends on the angle θ. Yet, it is often observed that θ is greater with humidification than with drying. In other words:

$$\left(\frac{1}{R_I}\right)_{hum} < \left(\frac{1}{R_I}\right)_{dry}$$

Thus:

$$P_{C\,hum.} < P_{C\,dry.}$$

This inequality provides one explanation for hysteresis. It is not the only explanation.

4.7.3. *Hysteresis and the ink-bottle model [EL 77]*

Consider an ink-bottle-shaped pore whose open end is constricted, i.e. having a "neck" that is open at both ends but smaller in diameter than the partially-open pore.

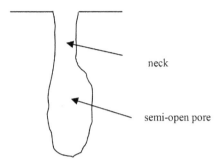

neck

semi-open pore

Figure 4.12. *Partially-open pore with constricted neck*

If the initially anhydrous solid is humidified with a gaseous mixture in which the water-vapor content is gradually increased, the water will condense on the walls of the pore, at a faster rate where the concave curvature of the walls is more pronounced. In other words, condensation will occur more quickly in the neck of the constricted pore so that it becomes filled with liquid before the partially-open pore. Consequently, the solid will exhibit retardation of humidification.

Conversely, if we dry the atmosphere in contact with a constricted pore initially filled with water, evaporation in the neck will begin, for the level of humidity corresponding to the curvature of the meniscus formed by the liquid filling the pore. When the neck is dry, evaporation will continue in the partially-open pore, on the walls whose curvature is less pronounced. However, when the humidity of the gaseous phase reaches a level that is intermediate between those at which equilibrium occurs in the neck and the partially-open pore, the entire structure remains filled with liquid and thus the solid exhibits drying retardation.

The function $\psi = f(R_o)$, which describes true thermodynamic equilibrium, is thus difficult to obtain. However, where drying is concerned, only the function which describes the dehydration is of practical use.

However, this explanation is less convincing than the inequality of the contact-angle values observed during humidification and drying (through imbibition and drainage).

NOTE.–

[MOR 65] experimentally examined the phenomena of imbibition and drainage in a porous medium.

They observed:

– the existence of a residual (irreducible) humidity which the authors referred to as bound humidity and which corresponds to a pendular state (saturation S_1);

– the existence of *gaseous globules* which prevent the completion of the saturation process. These globules only occur if imbibition takes place on a material which has been previously drained.

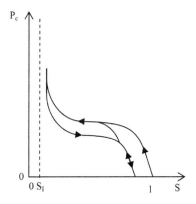

Figure 4.13. *Imbibition/drainage cycle (inorganic material)*

4.7.4. *Hysteresis in biological tissue*

[YOU 67] posited that a cell of living matter contains water from three distinct sources:

– strong adsorption as a molecular layer bound to the surface of the cell;

– adsorption through normal condensation occurring on part of the molecular layer;

– absorption within the cell via the monomolecular layer.

During the sorption process, since the first layer is bound to the surface, the molecules forming that layer cannot cross the wall. Thus, the absorbed quantity is proportional to the surface area covered by at least one layer produced by normal condensation on the monomolecular layer.

During the drying process, the absorbed humidity can only be completely removed if the bound layer no longer exists. Thus, any existing absorbed humidity will be proportional to the bound moisture – i.e. to the first layer.

The authors provide the analytical expressions for the sorption- and desorption isotherms.

In Figures 4.13 and 4.14, we see that the hysteresis loops for biological and inorganic matter are different. In biological matter, hysteresis is linked not to capillarity but to absorption within the cell.

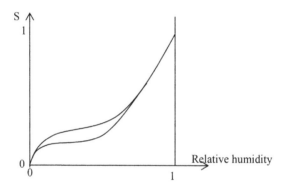

Figure 4.14. *Shape of the hysteresis loop for biological matter*

4.7.5. *Hysteresis and capillary condensation (according to [BRO 67, BRO 68])*

When a molecule undergoes condensation, its chemical potential changes from μ_V to μ_C.

In a cylindrical pore with a surface area A, the condensation of dn molecules will cause this surface area to change by dA:

$$Vdn = -(r-t)dA \quad \text{hence} \quad dA = -\frac{V}{r-t}dn$$

V: molar volume: $m^3 \cdot kmol^{-1}$

r: radius of the pore without condensation: m

t: thickness of the condensation layer: m

During condensation, the change in Gibbs energy is:

$$dG = (\mu_C - \mu_V)dn + \gamma dA = (\mu_C - \mu_V)dn - \frac{\gamma V}{r-t}dn$$

However, Broekhoff and de Boer hypothesize that the chemical potential μ_C of the condensate is modified as a result of the closeness of the wall, and that the extent of the modification can be obtained by subtracting a function F(t) of the thickness of the condensate from the chemical potential of the latter. This function of the thickness t is:

positive

decreasing

zero when $t = r$ (along the axis of the pore).

We can therefore write:

$$\frac{dG}{dn} = \mu_C - \mu_V - \frac{\gamma V}{r-t} - F(t)$$

Let us calculate the integral for t varying from t_e to r. The thickness t_e corresponds to $dG/dn = 0$. This is the equilibrium thickness.

$$\Delta G = \int_{t_e}^{r} \left[\mu_C - \mu_V - \frac{\gamma V}{r-t} - F(t) \right] 2\pi(r-t)dt$$

Indeed, $dn = 2\pi(r-t)dt$ for the unit length of the pore:

$$\Delta G = (\mu_C - \mu_V)\left(\frac{r-t_e}{2} \right)^2 - (r-t_e)\gamma V - \int_{t_e}^{r} (r-t)F(t)dt$$

ΔG is the variation in Gibbs energy corresponding to the condensation of a film of thickness $(r - t_e)$ – i.e. the condensation of Δn molecules.

For any given value of $(\mu_V - \mu_C)$, the shape of the changes in dG/dn is shown in Figure 4.15.

The function dG/dn has a maximum value for which $d^2G/dn^2 = 0$

$$\frac{d^2G}{dn^2} = \left[-\frac{\partial F}{\partial t} - \frac{\gamma V}{(r-t)^2} \right] \frac{dt}{dn} = 0 \text{ with } \frac{dt}{dn} \neq 0$$

This maximum corresponds to a critical value t_{cr} of t.

If we are looking for a stable equilibrium, d^2G/dn^2 must be positive, which requires that:

$0 < t < t_{cr}$

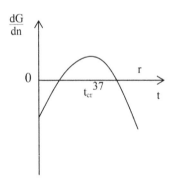

Figure 4.15. *Shape of the changes in dG/dn*

Finally, the changes in ΔG as a function of t are shown in Figure 4.16, with the following parameter:

$$(\mu_C - \mu_V) = RT \ln \frac{P_C}{P_V}$$

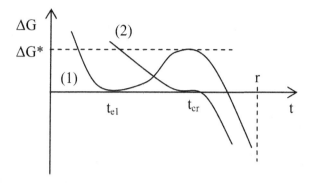

Figure 4.16. *Shape of the variations in ΔG*
as a function of the thickness t of the condensate

Thus, we can observe that, in general, and again for any given value of $\text{Ln}\,\dfrac{P_v}{P_c}$, there will be partial filling of the pore, characterized by the equilibrium thickness t_e, and that it is only with a specific value of $\dfrac{P_v}{P_c}$ that we will have both:

$$\frac{\partial \Delta G}{\partial n} = 0 \ \text{ and } \ \frac{\partial^2 \Delta G}{\partial n^2} = 0$$

At that point, the thickness t may increase to r without being prevented from doing so by the presence of a Gibbs energy barrier ΔG^*. This value of P_V/P_C corresponds to the *sudden filling of the pore*. This phenomenon is known as capillary condensation.

In Figure 4.16, we can see that, for curves (1) and (2), we have:

$$\left(\mu_c - \mu_v\right)_{(1)} > \left(\mu_c - \mu_v\right)_{(2)}$$

Put differently:

$$P_{V2} > P_{V1} \ \ \sin ce \ \ \mu = \mu_0 + RT \ln P$$

The vapor pressure must be increased in order to obtain capillary condensation. Thus:

– taking t_{e1} as a starting point, P_V must be increased from P_{V1} to P_{V2} in order to fill the pore by capillary condensation;

– taking the pore filled at P_{V2} as a starting point, the pressure must be lowered from P_{V2} to P_{V1} in order to empty it partially at t_{e1}.

This accurately describes the *phenomenon of hysteresis*.

4.8. Movements of the fluids

4.8.1. *Biphasic flow. Composition of the flow [LEV 41]*

For each of the two fluids, the percolation equation is as follows:

$$\frac{\mu_M q_M}{K_M} = -\left[\frac{dP_M}{dz} - \rho_M g \sin \alpha\right]$$

$$\frac{\mu_m q_m}{K_m} = -\left[\frac{dP_m}{dz} - \rho_m g \sin \alpha\right]$$

Subtracting the second equation from the first, term by term, gives us:

$$\frac{P_M q_M}{K_M} - \frac{\mu_m q_m}{K_m} = \frac{dP_c}{dz} - g \sin \alpha \Delta \rho = X \qquad [4.9]$$

q_M and q_m: volumetric flux densities of the two fluids: $m.s^{-1}$;

K_M and K_m: permeabilities for the two fluids: m^2;

P_M and P_m: pressures of the two fluids: Pa;

ρ_M and ρ_m: densities of the two fluids: $kg.m^{-3}$

$$\Delta \rho = \rho_M - \rho_m$$

μ_M and μ_m: viscosities of the two fluids: Pa.s;

α: angle of the direction z with the horizontal: rad;

g: acceleration due to gravity: $m.s^{-2}$.

Let us posit that:

$$q_M = f_M q_T; \quad q_m = (1-f_M)q_T \qquad [4.10]$$

Thus:

$$f_M = \frac{\dfrac{X}{q_T} + \dfrac{\mu_m}{K_m}}{\dfrac{\mu_M}{K_M} + \dfrac{\mu_m}{K_m}} \qquad [4.11]$$

Equations [4.9], [4.10] and [4.11] give q_T, f_M, q_M and q_m.

4.8.2. Relative permeabilities

Consider the case where a gas and a liquid must simultaneously percolate through a porous solid.

The limiting pore radius R is such that:

– only the gas flows through the pores having a radius greater than R;

– only the liquid flows through the pores having a radius less than R.

The limiting radius R corresponds to a level of saturation S defined as follows:

$$S = \frac{\text{volume occupied by the liquid}}{\text{total volume of the pores}}$$

According to Poiseuille's law, the flowrate of a fluid through a pore of radius r is:

$$q = \frac{\pi \Delta P r^4}{8 \mu L}$$

ΔP: pressure drop across the porous solid: Pa.s;

μ: viscosity of the fluid: Pa;

L: true length of the pore: m.

According to Laplace's law:

$$r^2 = \frac{4\gamma^2}{P_C^2}$$

Moreover, let us define a distribution law β (r) for pore cross-sections such that the total cross-section exposed to the fluids is:

$$A_T = \int_0^\infty \pi r^2 \beta(r)\, dr$$

The cross-section exposed to the liquid alone is:

$$A_L = \int_0^R \pi r^2 \beta(r)\, dr$$

The length L, considered to be common to pores of all sizes, is:

$$L = tZ$$

where t is the coefficient of tortuosity, and Z the thickness of the porous plate.

The saturation is therefore:

$$S = \frac{A_L}{A_T} = \int_0^R \pi r^2 \alpha(r)\, dr$$

where:

$$\alpha(r) = \frac{\beta(r)}{\int_0^\infty \pi r^2 \beta(r)\, dr}$$

Thus:

$$dS = \pi r^2 \alpha(r) dr$$

In light of the above, the total liquid flowrate should be:

$$Q_L = \frac{\pi \Delta P A_T}{8 \mu t Z} \int_0^R \frac{4\gamma^2}{P_C^2} dS$$

In fact, experience shows that this formula is incorrect.

At the inlet, the cross-section exposed to the fluids is $A_F = \varepsilon A_T$ where ε is the porosity. If we sweep the cross-section A_F, the probability of encountering the liquid is S. Likewise, at the outlet, the probability of encountering the liquid is also S, whereas the probability of encountering the gas is (1-S).

If we sweep the entire surface A_F, obviously, the probability of encountering one fluid or the other is 1.

However, we can write the identity:

$$1 = (1-S)^2 + (1-S)S + S(1-S) + S^2$$

The first term corresponds to a pathway where the gas is encountered both at the inlet (probability (1-S)) and at the outlet (probability (1-S)). The probability of the two events occurring simultaneously is $(1-S)^2$.

The second term denotes that the gas is encountered at the inlet and the liquid at the outlet. This scenario cannot be considered a flow, since the gas cannot be converted into liquid. The same is true for the third term.

The fourth term denotes the probability of encountering the liquid on both sides.

Consequently:

– the pore cross-section fraction allowing the passage of the liquid is S^2;

– the pore cross-section fraction allowing the passage of the gas is $(1-S)^2$.

These conclusions stem from the fact that the pores are interconnected in a random manner, and that a fluid subjected to a pressure gradient cannot flow unless it takes the form of continuous, uninterrupted streams. Throughout the drying process, these streams remain continuous since the number of streams and the size of their cross-sections decrease, but the cross-sections in question are precisely those of the pores which contain them. The number and cross-section of these pores also decrease as the porous substance releases its water.

The pores in question are those that are open to the outside world on each of the faces. Pores of the same diameter that are non-continuous are also involved, like the others, in the evaporation process, but are not involved in the migration of the liquid.

Finally, the liquid flowrate is:

$$Q_L = \frac{A_F \pi \Delta P \gamma^2 S^2}{2\mu t Z} \int_0^S \frac{dS}{P_C^2}$$

and the gas flowrate:

$$Q_G = \frac{A_F \pi \Delta P \gamma^2 (1-S)^2}{2\mu t Z} \int_S^1 \frac{dS}{P_C^2}$$

If only one fluid were involved, its flowrate would be:

$$Q_F = \frac{A_F \pi \Delta P \gamma^2}{2\mu t Z} \int_0^1 \frac{dS}{P_C^2}$$

The relative permeability of the liquid is therefore:

$$K_{RL} = \frac{Q_L}{Q_F} = \frac{S^2 \int_0^S \frac{dS}{P_C^2}}{\int_0^1 \frac{dS}{P_C^2}}$$

and that of the gas is:

$$K_{RG} = \frac{Q_G}{Q_F} = \frac{(1-S)^2 \int_S^1 \frac{dS}{P_C^2}}{\int_0^1 \frac{dS}{P_C^2}}$$

As we have seen, there is a simple relation between the capillary pressure P_C and the saturation of a porous medium:

$$S = \left(\frac{P_P}{P_C}\right)^\lambda \text{ or indeed: } P_C = P_E S^{-\frac{1}{\lambda}}$$

P_E is the pressure at which the liquid begins to be expelled from the medium by the gas. This is the inlet pressure (of the gas).

The relative permeabilities are immediately computed by the following integral:

$$K_{RL} = S^{\frac{2}{\lambda}+3} \text{ and } K_{RG} = (1-S)^2 \left(1 - S^{\frac{2}{\lambda}+1}\right)$$

Wakerman observed that the values of λ range from 3.8 to 10, and proposed that 5 be adopted as the universal value.

4.8.3. Drying of a rigid, inert, porous medium

[WHI 77a, WHI 77b] proposed a description of the drying process based on the transport equations for the gaseous phase, the liquid phase and the energy transported by both of these phases, as well as by conduction. The conservation equations for the two phases, as well as those for the diffusion and convection of the gas, and the thermodynamic equilibrium relationships, complete the systems.

[MUJ 80] applied the above-mentioned ideas to the initial drying phase, which he considered as isothermal.

[ECK 80] and [NEI 82] analyzed and simulated the simultaneous transport of heat and moisture in a porous body.

[WHI 83, WHI 84] divided the porous solid into three regions, and replaced the intermediate region with a jump in drying characteristics.

[ECK 86] applied the theory of steady-state drying to a 5 cm thick plank of wood and reported the values of the physical parameters of the operation.

[BEN 86] applied the theory of drying to a brick and reported its characteristics.

These last two studies demonstrate that it is possible to simulate the drying of a porous body using a stream of hot air without recourse to excessively-complex computations.

Certain properties are calculated using methods that are more recent than those mentioned in sections 4.8.1 and 4.8.2.

5

Dryers

5.1. General

5.1.1. *The various types of dryers*

The study of the kinetics of drying leads to a classification into:

– convection dryers, where the operation often lasts less than an hour;

– diffusion dryers, where the operation may last for several hours and, sometimes, several weeks.

However, the surface area available for the evaporation also plays an important role. In convection devices, such as pneumatic dryers, fluidized beds or atomization towers, each particle of product is individually in contact with the gas and the transfer of humidity is very quick.

On the other hand, in boxes and vats, and to a lesser extent, in stirred discontinuous-flow dryers and rotary drums, the motion of the product is little or non-existent, and the heat transfer may require several hours, because it takes place only at the free surface of the product, or else at points of contact with the heating surface.

Thus, dryers can be classified according to whether they operate by direct transfer direct (convection) or indirect transfer (conduction across a metal wall). Indirect transfer must be used when we wish to avoid the entrainment of dust or the contamination of the product by hot gases.

A single dryer may be incapable, on its own, of delivering the desired degree of dryness. In this case, we must combine several devices in a series.

Thus, an atomizer or a pneumatic dryer may be followed by a fluidized bed or a calcining kiln.

It may be that the product being processed contains a costly or hazardous solvent which must not be released into the atmosphere. In this case, drying is carried out in a closed circuit.

When the feed product is too damp, it is possible to recycle a portion of the outgoing dry product, injecting it back into the feed.

Let R represent the recycling rate, i.e. the ratio of the recycled outgoing flowrate to the flowrate removed (production).

The fraction of the total flowrate sampled is $x = 1/(1 + R)$.

The recycled fraction is $(1 - x)$.

After the second passage, we remove $x(1 - x)$ and recycle $(1 - x)^2$.

After the third passage, we remove $x(1 - x)^2$ and recycle $(1 - x)^3$.

After the nth passage, we remove $x(1 - x)^{n-1}$ and recycle $(1 - x)^n$, etc.

Thus, we can deduce the distribution of the lengths of stay in the outgoing product. This distribution is notably expanded in comparison to what it would be in the case of a single-pass dryer.

Drying devices can, finally, be classified into continuous-flow or discontinuous-flow machines. In the former case, up to 50 tons of product can be processed per hour, whereas, in the latter case, we can go no higher than 5 tons.h^{-1}.

5.1.2. *Phases and components*

When performing the necessary calculations for a dryer, we need to determine the temperature T_e of the air as it enters the device. It is always best to take the air as hot as possible, as this:

– decreases the flowrate of air, and therefore the price and power of the fans;

– decreases the necessary investment in general;

– increases the thermal yield of the installation.

The yield in question increases with T_e and is defined by:

$$\frac{T_e - T_s}{T_e - T_a}$$

T_e: inlet temperature of the air into the dryer,

T_s: outlet temperature of the air from the dryer,

T_a: ambient temperature, which is that of the air taken in and heated to T_e.

The hotter the outgoing air is, the more energy is wasted.

The temperature T_e is limited by the qualities of the working fluids available (vapor, electricity) and also by the product's heat resistance.

Having determined the temperature T_e, we now need to set the outlet temperature of the product. One possibility is to choose a temperature which is 10–40°C higher than the boiling point of the liquor impregnating the product. Another possibility is to choose a temperature 20–40°C lower than the air inlet temperature (in the case of countercurrent circulation).

The air outlet temperature must be 10–30°C higher than that of the incoming product (countercurrent circulation) or outgoing product (co-current circulation).

Of course, the inlet temperature of the product is fixed, so we simply need to write the heat balance to find the useable air flowrate (to which we often must add a further flowrate in order to take account of heat losses).

If the air outlet temperature is lower than 100°C, it is crucial to check that the air will not reach saturation with water vapor after a brief cooling period.

5.2. Batch dryers

5.2.1. *Advantage*

These devices entail little initial investment, they are easy to maintain and are simple to use. They are highly flexible in terms of their applications. A chamber in which tobacco leaves are hung to dry is a very economical solution.

5.2.2. *Vacuum dryer with scraper*

Such dryers are in the form of a cylinder with a vertical axis whose height is less than its diameter. The bottom and the walls are heated by vapor. An anchored stirrer sweeps the product. The device is loaded and unloaded at the top after being overturned.

The diameter of these devices varies from 1 to 3 m, and thus the useable volume may be up to 4 m^3. The stirrer turns at 2 to 20 rpm^{-1}. The mean heat transfer coefficient ranges from 60 to 90 $W.m^{-2}.°C^{-1}$ and increases with the humidity of the product.

The power consumed by the stirrer passes through an acute maximum which corresponds to a thick paste. During drying, the flow of evaporation also passes through a maximum. Beyond this, we obtain a divided solid.

This device is recommended for operations relating to small quantities of material. It is useful for products that are very dusty at the end of the drying process. In fact, the attrition of the dry product is significant.

5.2.3. *Fluidized-bed dryer*

In a fluidized bed, the order of magnitude of the drying time of crystals is between minutes and several hours for seeds.

The result of this is that the humidity of the gas varies little between the inlet and outlet, and for the calculations, this humidity can be assimilated to its value Y_0 at input.

This means that the product is submitted to an ambience with constant quality.

5.2.4. *Drying cupboards*

In these dryers, the product to be treated is laid in flat troughs or on superposed plates loaded by hand as regularly as possible to prevent local overheating, which is encouraged by the high conductivity of the bottom of the plates. The thickness of the layer of solid ranges from 1 to 5 cm. Certain devices are capable of treating up to $1m^3$ of product.

When we use grates whose bottom is made not of solid sheet metal but metal trellises or pierced sheet metal, the risk of overheating is significantly lesser. These boxes are different from vats because the former are loaded and emptied from the outside, whereas it is possible to enter into the latter.

The air circulates either at the surface of the plates, lapping at the product (horizontal ventilation), or through the product (transverse ventilation). In the latter disposition, the drying time is much shorter.

In any case, plates situated near to the air inlet will be subject to significantly different conditions to those existing near to the air outlet. This is problematic when dealing with products that are sensitive to heat. We can deal with this issue by removing the plates at different time intervals or, for instance, reversing the direction of the gas stream.

The drying air temperature ranges from 30°C to 200°C. It is limited by the product's sensitivity to heat.

In general, the air is recycled.

The air at the outlet is more humid in the case of recycling, which results in a heat saving, but at the cost of a longer drying time.

Generally, the specific gas flowrates are low so as to avoid entrainment. Discontinuous-flow dryers are often kept at a slight negative pressure so as not to pollute the atmosphere of the workshop.

All of this means that only the surface area of the plates is the main defining parameter.

In horizontal vent dryers, the evaporation rate is 0.15 to 1.5 $kg.h^{-1}m^{-2}$ of water. When the air is heated by vapor, we need 2.2 to 2.5 kg of vapor per kg of water evaporated.

In transverse vent dryers, the velocity of the air ranges from 0.5 to 1.5 m.s^{-1}. For a thickness of layer of up to 3cm, the pressure drop would reach 3 mbar. It is 50 mbar for a layer 50 cm in height. The evaporation rate reaches 12 kg.h^{-1}m^{-2}.

When the product to be dried is expensive and fragile, it is preferable to use a device where it remains immobile. Such is the case with boxes and vats. However, the production of these devices is limited.

A satisfactory device must have the following characteristics:

– the plates must be interchangeable,

– the interior must be easily accessed,

– the device must be easy to maintain and clean,

– there must be sufficient insulation so as not to cause discomfort to staff working around the device.

Vats are very widely used in the pharmaceutical industry. They are also used to dry wood, wool, etc.

5.3. Atomization tower and spherulation tower

5.3.1. *Principle of the method*

The product being treated needs to be a liquid. It is pulverized in a cylindrical chamber in which circulates a hot gas in a countercurrent direction to the droplets. Heat- and material transfer takes place by convection between the gas and the droplets.

At the outlet, the gas and the coarse particles can be recovered at the bottom of the atomization tower. The finer particles are recovered in a cyclone or a sleeve filter.

The inlet temperature of the gas varies, depending on the products at hand, from 150 to 800°C.

The evaporation capacity of atomization towers is significant – it can be as great as 15 tons.h^{-1}. The length of stay of the gas in the tower ranges from 4 to 10 s for towers up to 5 m in diameter and from 10 to 30 s or more with larger towers.

5.3.2. *Spraying systems*

1) Pressure sprayers

In these devices, the feed tube is followed by a turbulence chamber designed to endow the liquid with helicoidal motion. The liquid then escapes through an orifice whose diameter can range from 0.3 to 4 mm. The pressures used can range from 20 to 250 bar (viscous liquids).

Sprayer buses are vulnerable to incrustations, blockages and erosion.

A purely-indicative expression links the pressure drop to the mean drop diameter in the case of a liquid that is not overly viscous:

$$D_g = \frac{200}{P^{1/2}}$$

D_g: microns;

P: bar.

The Partiole size distribution (PSD) obtained depends heavily on the characteristics built in by the manufacturer. It is far better to read the manufacturer's manual than to employ any kind of general formula. We can, however, add that pressurized buses are used when we want to obtain particles whose diameter is smaller than 200 μm.

As the nozzle is designed for a given flowrate and PSD, its operation lacks flexibility.

In light of its drawbacks, this device is used only for small installations (around the size of a pilot).

2) Pneumatic sprayers

In this case, the liquid arrives through a central tube surrounded by a ring-shaped tube carrying air. In addition to mechanical pulverization, we have aeraulic pulverization due to pressurized air. Thus we can obtain extremely fine particles, of the order of 5–10 μm.

This process is used only for small installations (up to 400 kg.h^{-1} of air); it necessitates an increase in the dimensions of the chamber, because the introduction of cold air causes a drop in energy which must be compensated.

In these installations, the pressure of the liquid ranges from a very low value up to 4 bars. The air pressure ranges from 4 to 8 bar.

The mean diameter of the droplets produced by a pneumatic sprayer is given by the following expression:

$$D_g = 6.85 \times 10^5 \frac{V_L}{V_G} \left[\frac{\sigma}{\rho_L}\right]^{1/2} + 3300 \left[\frac{\mu_L}{\rho_L}\right]^{0.45} \left[\frac{1000\, Q_L}{Q_G}\right]^{1.5}$$

D_g: microns;

V_L and V_G: ejection rate of the liquid and of the gas: m.s^{-1};

σ: surface tension of the liquid: N.m^{-1};

ρ_L: density of the liquid: kg.m^{-3};

μ_L: viscosity of the liquid: Pa.s;

Q_L and Q_G: volumetric flowrates of the liquid and of the gas: m^3.s^{-1}.

3) Centrifugal sprayers

Centrifugal sprayers are hollow disks or "turbines" in which the liquid is accelerated and ejected from the rim at a peripheral velocity of 70–200 m.s^{-1}, with a typical value of 150 m.s^{-1}. The rotation speed of the mobile part ranges from 3000 to 20,000 rev.mn^{-1}, and its diameter is between 5 and 35 cm. A single sprayer can release 25–30 tons of liquid per hour.

Friedman, Gluckert and Marshall linked the harmonic mean droplet diameter to the operational parameters by the relation:

$$\frac{D_s}{R} = 0.4 \left[\frac{\Gamma}{\rho_L N R^2}\right]^{0.6} \left[\frac{\mu}{\Gamma}\right]^{0.2} \left[\frac{\sigma \rho_L P}{\Gamma^2}\right]^{0.1}$$

Here:

D_s: harmonic mean diameter, which is the diameter of the droplet that has the same surface-to-volume ratio as the population in question;

R: radius of the disk: m;

Γ: line load in relation to the wet perimeter of the disk: $kg.m^{-1}.s^{-1}$;

ρ_L: density of the liquid: $kg.m^{-3}$;

N: number of turns per second;

μ: viscosity of the liquid: Pa.s;

P: wet perimeter of the disk: m;

σ: surface tension of the liquid: $N.m^{-1}$.

"Turbines" can operate very flexibly.

The dimension of the grains obtained is highly uniform, with a log-normal distribution.

The droplets are moving at high velocity when they leave the edge of the disk, and are subjected to energetic braking when they come into contact with the air. The duration of that braking is very slight: around 3/1000 of a second. During that time, evaporation is intense and takes place at the humid temperature of the air.

Disks and turbines are reliable. They can spin for 10,000 hours with no need for any intervention.

However, it is important to be aware of the presence of critical velocities – which is always a possibility – and also to pay attention to the corresponding vibration phenomena.

5.3.3. Design and advantage of towers

The product is usually fed into the device by volumetric pumps, and more specifically, membrane pumps for suspensions.

The shape of the towers depends on the sprayer used (see Figures 5.1 and 5.2).

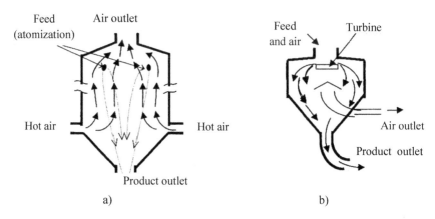

Figure 5.1. *a) Countercurrent spherulation tower (tall and narrow), b) atomization tower (co-current)*

Counter-current towers are narrow and tall.

In the case of pulverization by turbine, we use a cylindrico-conical tower that is almost as wide as it is tall. The reason for this is that the particles must not be allowed to reach the wall before they are dry, otherwise a deposit would form. The conical shape of the bottom of the tower prevents the stagnation of the product in a hot atmosphere.

Towers have the advantage of often producing spherical particles – often, but not always. Spherical particles flow easily.

It is possible to obtain a high apparent density for the product by avoiding sharp gradients of humidity and temperature within the particles as they dry. We do this by adopting a moderate temperature for the air at input and circulating the air in a counter-current direction.

We can evaluate the volume of an atomization tower by accepting a heat-transfer coefficient equal to 46 $W.m.^{-3}°C^{-1}$ and oversizing the volume calculated by a percentage proportional to that volume and equal to 50% for 2000 m^3.

EXAMPLE 5.1.–

Drying of an organic compound in an atomization tower.

We wish to dry 21,000 kg.h^{-1} of product at 25% D.E (dry extract) to bring it to 95% D.E.

The mass flowrate at output is $\dfrac{21,000}{3600} \times \dfrac{0.25}{0.95} = 1.54$ kg.s^{-1}

Thus, we have evaporated $\dfrac{21,000}{3600} - 1.54 = 4.29$ kg.s^{-1}

We use incoming air at 500°C and outgoing air at 120°C.

The product to be dried enters at 45°C and exits at 90°C. Its specific heat capacity is 3340 J.kg.$^{-1}$°C^{-1}.

The heat lost by the air serves to:

– heat up the product:

$Q_1 = 1.54 \times 3340 \times 45 = 231,462$ W

– evaporate the water:

$Q_2 = 4.29 \times 2,466,000 = 10,580,000$ W

Thus, in total: Q = 10,811,462 W.

The temperature program allows us to determine the log mean temperature difference (LMTD). We shall use the quick equivalent-exchanger method (see section 4.2.12).

$$
\begin{array}{ccc}
500 & \longrightarrow & 120 \\
45 & \longrightarrow & 90 \\
\hline
455 & & 30
\end{array}
\qquad \Delta T = 155°C
$$

With a heat transfer coefficient equal to 46 $W.m^{-3}.°C^{-1}$, the volume calculated for the chamber is:

$$V_c = \frac{10,811,462}{46 \times 155} = 1516 \text{ m}^3$$

The oversizing is:

$$50\% \times \frac{1516}{2000} = 38\%$$

Acceptable volume:

$$V_a = 1516 \times 1.38 = 2090 \text{ m}^3$$

EXAMPLE 5.2.–

Thermal pre-calculations for a counter-current tower.

The air is taken in at 20°C and 60% relative humidity, which corresponds to the partial pressure *at 1 atmosphere*.

$$P_V = 0.6 \times 0.024 = 0.0144 \text{ atm}$$

$$Y_{inlet} = \frac{By}{1-y} = \frac{18 \times 0.0144}{29(1-0.0144)} = 0.0091$$

B is the ratio $18/29 = 0.62$.

The flowrate of dry air is 6.6 $kg.s^{-1}$ and the tower needs to evaporate 0.12 $kg.s^{-1}$ of water. At its outlet, the air contains:

$$Y_{outlet} = 0.0091 + \frac{0.12}{6.6} = 0.0273 \text{ kg of vapor per kg of dry air}$$

$$y_{outlet} = \frac{0.0273}{0.62 + 0.0273} = 0.0422$$

Evaporation commences when the temperature of the product has risen to around the humid temperature of the outgoing air (see Chapter 1).

We shall accept a gap of 20°C between the temperature of outgoing air and its humid temperature. Supposing that the air is at 40°C, let us begin by calculating:

$$C_G \rho_G = 1000 \times 1.29 \times \frac{273}{313} = 1201 \text{ J.m}^{-3}.°C^{-1}$$

$$c_T = \frac{P}{RT} = \frac{10^5}{8314 \times 313} = 0.038 \text{ kmol.m}^{-3}$$

$$L M_E = \Lambda = 2.4 10^6 \times 18 = 43.2 \times 10^6 \text{ J.kmol}^{-1}$$

The psychrometric coefficient σ is equal to 1 for water vapor in air. Therefore we can apply equation [1.7], to the air outlet (taking $a_E = 0.8$):

$$t_G - t_H = \frac{L M_E c_T \sigma}{C_G \rho_G} \left(a_E \pi (t_H) - y_G \right)$$

$$20°C = \frac{43.2 \times 10^6 \times 0.038 \times 1}{1201} \left(0.8 \pi (t_H) - 0.0422 \right)$$

$$\pi (t_H) = 0.071 \text{ atm}$$

At the outlet: $t_H = 39.5°C$

The outlet temperature of the gas, therefore, is:

$$39.5 + 20 = 59.5°C$$

If we overlook the sensible heat of the product, the cooling of the air is:

$$\frac{0.121 \times 2.4 \times 10^6}{6.6 \times 1000} = 44°C$$

As the air exits at 59.5°C, it must enter at:

$$59.5 + 44 = 106.5°C$$

$$C_G \rho_G = 1000 \times 1.29 \times \frac{273}{376.5} = 935 \; \text{J.m}^{-3}.^{\circ}\text{C}^{-1}$$

$$c_T = \frac{10^5}{8314 \times 376.5} = 0.0319 \; \text{kmol.m}^{-3}$$

The humid temperature of the outgoing product is such that equation [1.7] is satisfied:

$$106.5 - t_H = \frac{43.2 \times 10^6 \times 0.0319}{935} \left(0.8\pi(t_H) - 0.0144 \right)$$

The equation of the humid temperature is satisfied by $t_H = 38^{\circ}\text{C}$. Hence, the product exits a little over 38°C after slight overheating. It enters at 20°C and is preheated to the humid temperature of 39.5°C. The temperature of the product varies very little.

The specific heat capacity of the product is $2100 \; \text{J.kg}^{-1}.^{\circ}\text{C}^{-1}$. Its humidity at input is $X = 0.4$.

The flowrate of dry product is: $0.12/0.4 = 0.3 \; \text{kg.s}^{-1}$.

The thermal powers are:

– Inlet of the product: $0.3 \times (2100 \times 1 + 0.4 \times 4180) (39.5 - 20) = 22.7 \times 10^3 \; \text{W}$.

– Outlet of the product à 60°: $0.3 \times 2100 (60 - 38) = 13.9 \times 10^3 \; \text{W}$.

The powers correspond to the following degrees of cooling for the air:

$$\frac{22.7}{6.6} = 3.44^{\circ}\text{C} \text{ and } \frac{13.9}{6.6} = 2.1^{\circ}\text{C}$$

It is therefore possible to plot the enthalpy diagram for the operation, in the knowledge that the thermal power of evaporation is:

$$0.12 \times 2.4 \times 10^6 = 290.4 \times 10^3 \; \text{W}$$

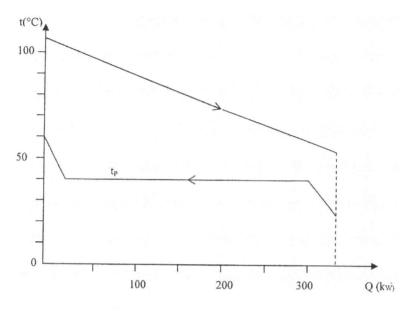

Figure 5.2. *Enthalpy diagram for an atomization tower(or spherulation tower)*

5.3.4. *Calculation for a spherulation column*

1) Heat balance

The mass flowrate W_G of gas needed to process the flowrate W_s of dry solid is given by:

$$\frac{W_G}{W_s} = K = \frac{C_s(t_{so} - t_{en}) + X(H_{so} - h_{en})}{C_G(T_{en} - T_{so})}$$

X: mass of water per mass of dry solid;

C_s and C_G: specific heat capacities of the dry solid and of the gas: $J.kg^{-1}.°C^{-1}$;

h_{in} and H_{out}: mass enthalpies of the liquid water at inlet and of the water vapor at outlet: $J.kg^{-1}$;

t_{in} and t_{out}: temperatures of the product at inlet and at outlet: °C;

T_{in} and T_{out} : temperatures of the gas at inlet and at outlet: °C.

2) Relative velocity between gas and product. Section of the column

This velocity is given by Richardson and Zaki (see sections 3.1.2 and 3.1.5 of [DUR 16d]). It is identical to the free-fall velocity of the particles.

By definition, the velocity V_R is given by:

$$V_R = \frac{U_P}{\beta} + \frac{U_G}{1-\beta}$$

U_p and U_G: velocities in an empty bed of the solid and of the gas: $m.s^{-1}$;

β: compacity (volumetric fraction) of the solid phase in the column.

However, for each of the two phases, we have:

$$U_P = \frac{W_s}{\rho_s A} \quad \text{and} \quad U_G = \frac{W_G}{\rho_G A}$$

ρ_S and ρ_G: densities of the particles and of the gas: $kg.m^{-3}$;

A: section of the column: m^2;

W_P and W_G: mass flowrates of the particles and of the gas: $kg.s^{-1}$.

Finally, the section (area) A of the column is, for a flowrate W_p of particles:

$$A = \frac{W_P}{V_1}\left[\frac{1}{\rho_P \beta} + \frac{K}{\rho_G (1-\beta)}\right]$$

For compacity values β of less than or equal to 0.03, we can accept that the limiting velocity of the particle cloud is assimilatable to that of an isolated particle, and is equal to the relative velocity V_R of the particles with respect to the gas.

For compacity values β greater than 0.03, we must use the value given in section 3.1.5 in [DUR 16d].

3) Exchange surface area: over a height Δh:

$$\Delta S = A\Delta h\beta \times \frac{6}{d_p}$$

d_p: particle diameter: m.

4) Heat transfer coefficient:

$$\alpha = \frac{\lambda}{d_p}\left[2 + 0.6\ Re^{0.5}Pr^{0.33}\right]$$

Let μ_G represent the viscosity of the gas (measured in Pa.s). We have:

Re: Reynolds number

$$Re = \frac{V_R d_p \rho_G}{\mu_G}$$

Pr: Prandtl number

$$Pr = \frac{C_G \mu_G}{\lambda}\ (\text{for air}, Pr \text{ is equal to } 0.7)$$

λ: conductivity of air: $J.s^{-1}.m^{-1}.°C^{-1}$

5) Thermal power transferred over a height Δh

$$\Delta Q = \alpha \Delta S \left(T_G - t_P\right)$$

NOTE.–

The above is applicable to solvent-based extraction using a so-called empty column, i.e. with no packing and no mechanical devices (such as disks or palettes) inside the column.

Note that flooding of the extractor is obtained for a limiting value of the compacity β (volume fraction of the dispersed phase). This value is significantly higher than 0.03.

5.4. Indirect-heating rotary dryer

5.4.1. *Description and advantage*

In this type of device, the heat necessary to vaporize the water is not provided by a hot gas, but instead by vapor tubes inside the ferrule. Thus, the gas flowrate can be significantly reduced, and may serve only to evacuate the water vapor that is formed.

The indirect-heating dryer is appropriate for fine products whose size is less than 100 μm. Indeed, as the velocity of the gas is low, the risks of entrainment are significantly reduced.

5.4.2. *Preliminary calculations*

Having determined a specific diameter for the dryer, we then need to deduce the mass flux density ϕ_G of the gas for which the diameter of the particles entrained is no higher than the pre-set value d_{pE}.

$$\phi_G = 4\sqrt{d_{pE}\rho_S\rho_G}$$

ϕ_G : mass flux density of the gas: $kg.m^{-2}.s^{-1}$;

d_{pE}: limiting sizing of particles re-entrained: m;

ρ_S and ρ_G: densities of the solid (with no interstitial spaces) and of the gas: $kg.m^{-3}$.

From this, we deduce the flowrate of the gas:

$$G = \phi_G \frac{\pi}{4}D^2 \qquad\qquad \left(kg.s^{-1}\right)$$

D: internal diameter of the ferrule: m.

The heat transfer coefficient (between the product and the gas) expressed per unit volume of the dryer is:

$$\alpha_v = \frac{325\phi_G^{0.16}}{D} \qquad\qquad \left(W.m.^{-2}{}^{\circ}C^{-1}\right)$$

This coefficient is of no use in the calculations for the device. On the other hand, it is crucial to know the corresponding material transfer coefficient (between the product and the gas), which is given by the Chilton–Colburn analogy. For water vapor, the psychrometric coefficient is equal to 1. Hence:

$$\beta_v = \frac{\alpha_v}{C_G \rho_G}$$

β_v: material transfer coefficient: $m.s^{-1}$;

C_G: specific heat capacity of the gas: $J.kg.^{-1\circ}C^{-1}$.

Generally, we know the boiling retardation of the saturated solution coating the crystals. Therefore, we can deduce the activity of the water:

$$a_e = \frac{\pi(100)}{\pi(100 + R_E)}$$

$\pi(t)$: vapor pressure of water at $t°C$: Pa;

R_E: boiling retardation: °C.

Having set the number and external diameter of the vapor tubes, we know the heat transfer surface S_u per meter length of the ferrule.

Similarly, we know the volume V_u of the dryer per meter length of the ferrule. It is the section $\frac{\pi}{4} D^2$ of the dryer.

5.5. Pneumatic dryer

5.5.1. Description

A pneumatic dryer consists of a vertical tube 10–20 meters long in which a hot gas circulates. The gas moves at a velocity between 12 and 25 $m.s^{-1}$, and that velocity is sufficient to entrain a solid in the state of dilute suspension.

The solid is introduced at the neck of a convergent/divergent vessel, using a screw or vibrating conveyer or indeed a rotary feeder valve. The ratio between the sections of the neck and of the drying tube is between 0.35 and 0.45.

If the feed is too humid, we can, like in other types of dryers, include recirculation of the dry solid by incorporating it into the feed with a paddle mixer (a trough in which there is a horizontal stick with paddles inclined on the axis, which homogenize the solid and move it along).

If a mill mechanism is put in place before the dryer, it is possible to sweep the inside of it with hot air, which partially dries the particles and facilitates milling by making the product less sticky. In this arrangement, two thirds of the drying can be performed in the mill before introduction into the dryer itself.

The ventilation system of a pneumatic dryer must overcome pressure drops across:

– the filter for the incoming air (cleanliness of the air);

– the array which heats the air;

– the dryer;

– the fine-particle separator generally composed of a cyclone, followed by a sleeve filter or a wet scrubber.

Often, therefore, it is necessary to have not one but two fans in place: one at the inlet after the filter and the other at the outlet from the washer.

5.5.2. Simulation in the divergent section

Calculations and experience both show us that drying takes place almost exclusively at the inlet of the device. In particular, at the neck of the divergent section, the velocity of the gas is two or three times greater than it is in the tube, and the product, which is initially immobile, is projected out of the divergent section and its velocity then decreases. As we shall see, in this initial phase, the difference between the velocities of the gas and the solid is significant, and drying is quick.

Thus, we divide the divergent section into N_T segments of equal length Δx. The segment with index i accommodates the section H_i at outlet; the section H_{i-1} at inlet; and, in the middle, the section $\overline{H}_i$. The volume of the segment i is ΔV_i.

We shall now present, for a segment:

– the preliminary calculations needing to be carried out;

– the thermal calculation.

5.5.3. *Preliminary calculations*

The force exerted by the gas on a particle is:

$$F_P = \frac{\pi d_p^2 C_X}{4} \cdot \frac{\rho_G V_R^2}{2}$$

V_R: relative velocity of the gas with respect to the particle: m.s^{-1}

$$V_R = |V_G - V_P|$$

d_P: diameter of the particle: m;

ρ_G: density of the gas: kg.m^{-3}

$$\rho_G = \frac{29}{22.4} \times \frac{273}{273 + t_G} \qquad \text{(pressure near to ambient)}$$

t_G: temperature of the gas: °C;

C_X: drag coefficient:

$$C_X = \frac{18.5}{Re^{0.6}}$$

Indeed, the flow regime is intermediate between the laminar and turbulent regimes.

Re is the Reynolds number for the particle:

$$Re = \frac{V_R d_p \rho_G}{\mu_G}$$

μ_G: viscosity of the gas: Pa.s.

The acceleration imparted to the particle is the quotient of the force of drag by the mass of the particle:

$$x'' = \frac{F_p}{m_p} = \frac{3 C_x \rho_G V_R^2}{4 \rho_s d_p}$$

ρ_s: true density of the solid: kg.m^{-3}.

In light of the expressions of C_x and Re, we obtain:

$$x'' = \frac{dx'}{d\tau} = \Gamma V_R^{1.4}$$

where:

$$\Gamma = \frac{13.87 \mu_G^{0.6} \rho_G^{0.4}}{d_p^{1.6} \rho_S}$$

Hence:

$$\Delta\tau_i = \frac{\Delta x_i'}{\Gamma \left| \overline{V}_{Gi} - \overline{x}_i' \right|^{1.4}} \qquad [5.1]$$

$\overline{V}_{Gi}$ and $\overline{x}_i'$ are overlined because they are expressed in the middle of the segment:

$$\overline{V}_{Gi} = \frac{G}{\rho_G \overline{A}_i}$$

G: mass flowrate of the gas: kg.s^{-1}.

$$\bar{x}_i' = x_{i-1}' + \frac{\Delta x'}{2}$$

Therefore, we must have:

$$\bar{x}_i' \Delta \tau_i = \Delta x$$

By successive iterations, we find the value of $\Delta x_i'$ so that the equation is satisfied, because we have set the value of Δx.

We then calculate:

1) The hold-up of the solid (volume fraction):

$$\phi_i = \frac{P}{\bar{x}_i' \bar{A}_i \rho_S}$$

P: mass flowrate of product: kg.s^{-1}.

2) The volumetric area for evaporation:

$$a_i = \frac{6\phi_i}{d_P}$$

3) The molar concentration:

$$c_{Ti} = \frac{273}{22.4(273 + \bar{t}_{Gi})} \left(kmol.m^{-3} \right)$$

As the mean temperature of the gas in the segment is unknown until we run the thermal calculations, we begin by taking:

$$t_{Gi} = t_{G,i-1}$$

$$\rho_{Gi} = 29\, C_{Ti}$$

4) The evaporation surface area:

$$\Delta S_i = a_i \Delta V_i$$

5) The Nusselt number:

$$Nu = 2 + 0.6 \ Re^{0.6} Pr^{0.33}$$

For air, the Prandtl number Pr is equal to 0.7.

6) The heat transfer coefficient:

$$\alpha = \frac{Nu \ \lambda_G}{d_P}$$

λ_G: heat conductivity of the gas: $W.m.^{-2\circ}C^{-1}$.

7) The flowrate of water vapor at the outlet from the segment with index $i - 1$:

$$E_{i-1} = G \ T_0 + \sum_{1}^{i-1} \Delta E_j$$

ΔE_j: flowrate of water evaporated in the segment with index j;

Y_0: absolute humidity of the incoming gas (kg of vapor per kg of dry air).

8) The mean absolute humidity of the gas in the segment:

$$\overline{Y}_i = Y_{i-1} + \frac{\Delta E_i}{2G}$$

Unless we actually run the thermal calculations, the value of ΔE_i is unknown. To begin with, we take:

$$\Delta E_i^{(o)} = 0$$

9) The mean specific heat capacity of the gas:

$$\overline{C}_{Gi} = C_G + \overline{Y}_i C_V \qquad \text{(see section 34.2.9)}$$

C_G and C_V are the specific heat capacities of the anhydrous gas and of water vapor.

10) The material transfer coefficient:

$$\beta_i = \frac{\alpha_i}{\overline{C}_{Gi}\rho_{Gi}}$$

11) The mean humidity of the product in the segment:

$$\overline{X}_i = X_0 + \frac{\sum_{j=1}^{i-1}\Delta E_j + \Delta E_i/2}{P}$$

12) The mean specific heat capacity of the product:

$$\overline{C}_P = C_P + \overline{X}_i C_e$$

C_P and C_e are the specific heat capacities of the anhydrous product and of water (in the liquid state).

5.5.4. *Thermal calculation*

We have the follow equations at our disposal:

1) The equation expressing that the heat lost by the gas has been transferred across the evaporating surface:

$$-G\overline{C}_G\left(t_{Gi} - t_{Gi-1}\right) = \alpha_i \Delta S_i \left[\left(\frac{t_{G,i-1} + t_{Gi}}{2}\right) - \overline{t}_{Pi}\right] \qquad [5.2]$$

$\overline{t}_{Pi}$ is the mean temperature of the product in the segment:

$$\overline{t}_{Pi} = t_{P,i-1} + \frac{\Delta t_{Pi}}{2}$$

2) The equation expressing that the heat lost by the gas has been used to heat the product and evaporate the water:

$$-G\overline{C}_G\left(t_{Gi} - t_{G,i-1}\right) = P\overline{C}_P\Delta t_{Pi} + L\Delta E_i \qquad [5.3]$$

3) The expression of the mean flowrate of water vapor:

$$\overline{E}_i = E_{i-1} + \frac{\Delta E_i}{2} \qquad [5.4]$$

4) The expression of the molar fraction of water vapor in the gas:

$$\overline{y}_{Gi} = \frac{\overline{E}_i / 18}{\dfrac{\overline{E}_i}{18} + \dfrac{G}{29}}$$

[5.5]

5) The expression of the flowrate of evaporated water:

$$\Delta E_i = M_E C_{Ti} \beta_i \Delta S_i \left(\overline{y}_{Pi} - \overline{y}_{Gi} \right)$$

[5.6]

6) The expression of the molar fraction of water vapor in a gas at equilibrium with the product:

$$\overline{y}_{Pi} = a_E \pi \left(\overline{t}_{Pi} \right)$$

[5.7]

a_E: activity of water;

$\overline{t}_{Pi}$ is the vapor pressure of water which appears at the temperature $\overline{t}_{Pi}$.

7) The expression of the variation of the temperature of the product:

$$\Delta t_{Pi} = 2 \left(\overline{t}_{Pi} - \overline{t}_{P,i-1} \right)$$

[5.8]

The approach to be followed consists of making a hypothesis about Δt_{Pi}. We then use the previous equations when we know $\Delta t_{Pi}^{(0)}$:

[5.2]	t_{Gi}
[5.3]	ΔE_i
[5.4]	$\overline{E}_i$
[5.5]	$\overline{y}_{Gi}$
[5.6]	$\overline{y}_{Pi}$
[5.7]	$\overline{t}_{Pi}$
[5.8]	Δt_{Pi}

We then take:

$$\Delta t_{Pi}^{(2)} = \frac{2}{3} \Delta t_{Pi}^{(0)} + \frac{1}{3} \Delta t_{Pi}^{(1)}$$

The convergence is quick.

5.5.5. *Calculation in the tube*

As regards the velocity of the solid, the preliminary calculations are simplified, because it is possible, *a priori*, to impose a variation of velocity Δx' and directly deduce the length Δx traveled by the product. The velocity profile of that product as it travels along the tube is therefore easily obtained.

When we know this profile, we merely need to divide the tube into sections and apply the above procedure to each of those sections. It would be prudent to give the length Δx, common to all those sections, a value no greater than 0.1 m. Indeed, as drying has practically completed by the time the mixture exits the divergent, little water remains to be evaporated (it is supposed that the length of the divergent is approximately 1 meter).

5.5.6. *Superheating of the product*

Once evaporation has completely finished, we need to superheat the product. For this purpose, we accept a final temperature gap between the gas and the solid equal to only 0.5°C.

The gas–solid contact surface necessary for this operation is then:

$$\Delta S = \frac{Q}{\alpha \Delta T} \qquad \qquad \left(m^2 \right)$$

ΔT: log mean temperature difference (LMTD): °C;

Q: thermal power corresponding to superheating: W;

α: heat transfer coefficient: $W.m.^{-2}°C^{-1}$.

The volume of the tube can be deduced from this:

$$\Delta V = \frac{\Delta S}{a_L} \qquad \qquad \left(m^3 \right)$$

and the length:

a_L: volumetric surface of the liquid: m^{-1}

$$\Delta x = \frac{\Delta V}{A} \qquad (m)$$

A is the section area of the tube: (m^2)

5.5.7. Section of the neck of the Venturi

At the inlet to the Venturi, the overpressure ΔP_V of the air with respect to the surrounding atmosphere is equal to the driving pressure of the ventilation fan, less the pressure drops across the air filter and the heating array.

The cross-section of the neck must be such that this overpressure is transformed into kinetic energy, so the pressure becomes identical to the external pressure. Let V_G and V_{GC} represent the velocities of the gas in the tube and at the neck of the Venturi. Bernouilli's theorem is written:

$$\frac{1}{2}\rho_G V_G^2 + \Delta P_V = \frac{1}{2}\rho_G V_{GC}^2$$

Let σ be the ratio of the cross-section of the neck to that of the tube. We obtain:

$$\Delta P_V = \frac{1}{2}\rho_G V_G^2\left[\frac{1}{\sigma^2} - 1\right]$$

EXAMPLE 5.3.–

$$\Delta P_V = 1500 \text{ Pa} \qquad \rho_G = 1 \text{ kg.m}^{-3} \qquad V_G = 20 \text{ m.s}^{-1}$$

$$1500 = \frac{1 \times 20^2}{2}\left[\frac{1}{\sigma^2} - 1\right]$$

$$\sigma = 0.343$$

5.5.8. *Air inlet to the Venturi*

It is prudent to take account of an inlet of ambient air which is always possible at the neck of the Venturi. Thus, the gaseous flowrate in the dryer may be increased by 15 to 20%, and the temperature of the resulting mixture is therefore lower than that of the air exiting the heating array.

5.5.9. *Practical data and use*

Generally, the ratio of the gas flowrate to the solid flowrate (expressed in terms of mass) ranges from 1.5 to 5 depending on the humidity of the feed (which varies from 0.03 to 0.50).

Products which stick to the walls are to be avoided, but it is possible to remedy this problem by recycling a portion of the outgoing dry product.

When the harmonic mean particle size is less than 150 μm, pneumatic dryers should not be used (and nor should any air-swept dryer), because it would cause difficulty in separating the gas and the product once it is dry.

In addition, shock-sensitive products should not be exposed to a pneumatic dryer.

On the other hand, as the contact between the product and the gas lasts for around one second, and given the co-current circulation of the gas and the solid, it is possible to dry thermosensitive products.

As drying occurs quickly, this device is suitable only for easy products – i.e. products whose particles are simply coated (rather than impregnated) with liquid, and for which the activity of the water in the coating liquor is greater than 0.4.

The inlet temperature of the gas ranges from 150 at 700°C and, up to 180°C, the gas can be vapor-heated. Beyond that temperature, we need electrical heating or combustion gases.

The pressure drop in the air on passing through the whole installation is between 200 and 500 mmH$_2$O.

5.5.10. *Quick estimation of a device*

We calculate the thermal power by accepting that, at the outlet, the air temperature is 10°C higher than that of the product. The thermal power is, of course, the sum of:

– the sensible heat of warming of the product;

– the latent heat of evaporation of the water.

If we take a heat transfer coefficient of approx. 2300 $W.m^{-3}°C$ and use the LMTD, we can deduce the volume of the drying tube. This is an application of the equivalent-exchanger method (see sections 4.2.11 and 4.2.12 of this book).

EXAMPLE 5.4.–

Suppose we want to dry 0.472 $kg.s^{-1}$ of wet dye at 20°C to increase the concentration of its dry extract from 70% to 99%.

The product is treated with 2.67 $kg.s^{-1}$ of air entering at 250°C and exiting at 100°C.

The diameter of the tube is 500 mm and its length is 10m.

Find the temperature of the product at the outlet, knowing that the specific heat capacity of the dry product is 1,463 $J.kg.^{-1}.°C^{-1}$.

We calculate the heat transferred by accepting a temperature of 90°C for the product at the outlet.

Quantity of water evaporated:

$$0.472\left[(1-0.7)-\frac{(1-0.99)\times0.7}{0.99}\right]=0.137 \ kg.s^{-1}$$

Heat for evaporation:

$$Q_e = 0.1377\times2,592,000 = 355,104 \ W$$

Sensible heat of the dry product:

$$0.472 \times 0.7 \times 1463 \times (90 - 20) = 33,836 \text{ W}$$

Thus, in total:

$$Q = 388,940 \text{ W}$$

The transfer takes place in a tube whose volume is:

$$V = \frac{\pi}{4} \times (0.5)^2 \times 10 = 1.96 \text{ m}^3$$

Hence, we find the log mean temperature difference:

$$\Delta T = \frac{388,940}{1.96 \times 2300} = 86.2°C$$

which corresponds to the following thermal program:

| 250 | $\longrightarrow$ | 100 |
| 20 | $\longrightarrow$ | 80 |
|-----| |-----|
| 230 | | 20 |

The product exits the dryer at 80°C.

5.6. Cylinder dryers

5.6.1. *Principle and description*

These devices comprise a hot metal surface over which the product is stretched. The surface may be flat (in the case of a conveyor belt) or cylindrical. It is the cylindrical device which we shall examine here.

The cylinder dryer can be used for initially-syrupy products. Such products stick to the metal surface and may form a very thin film whose thickness ranges from 80 μm to 600 μm.

The adherence of the film to the cylinder depends on the surface tension, the viscosity of the product and, of course, its granular nature (PSD).

Thus, we obtain direct contact between the heating wall and the product, which is economical in terms of energy, because the heat is not entrained and wasted by the air, which would still be hot on exiting. Indeed, the cylinder is heated by vapor.

In addition, the metal wall of the cylinder is thick. Its thickness may range from 2 to 4 cm, so the temperature of the wall in contact with the product is lower than the temperature of the face on which the heating vapor condenses. This means the device can be used to treat thermosensitive products – particularly when the rotation period of the device is short, between 2 and 15 seconds. Of course, the speed of rotation of the cylinder falls if we increase the thickness of the film and if the humidity of the product being treated is high. Hence, materials which are sensitive to temperature must be spread out in a thin layer. The flux density of vaporization is of the order of 25 $kg.h^{-1}.m^{-2}$.

The simplest means of loading is to partially immerse the cylinder in a stirred trough. The thickness of the layer obtained depends on the immersion, on the surface temperature of the cylinder, on its rotation speed and on the nature of the product.

If the properties of the feed are not absolutely constant, we can nevertheless obtain a thin, regular layer by intercalating one or two rollers between the tank and the drying cylinder. These injection rollers push the product through the tank and transport it to the drying cylinder, to which they are applied with a precise amount of pressure.

Materials which would be difficult to entrain by direct contact with the cylinder can be applied in the form of a mist projected by a brush roller which, instead of hairs, has metal spurs.

Besides single-cylinder devices, dryers containing two cylinders have been designed.

1) Upward cylinders (that is, turning upward in relation to the feed).

a) dipping cylinder

Figure 5.3. *Means of loading*

The direction of rotation of the cylinders is shown in Figure 5.4. The feed comes in at the top, between the cylinders.

When the feed is very humid or when it contains rough and seedy particles, we must choose the twin-cylinder dryer, whose production capacity is obviously twice that of a single-cylinder device.

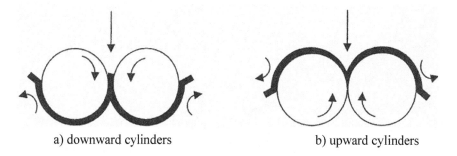

a) downward cylinders   b) upward cylinders

Figure 5.4. *Twin-cylinder dryers*

2) Downward cylinders (that is, turning downward in relation to the feed).

In this type of dryer, the product is fed in between the two cylinders. The gap left free between the cylinders determines the thickness of the two films obtained, independently of the properties of the product being treated. However, the proportion of the surface used is low, as is shown by Figure 5.4.

5.6.2. *Heat flux density for vaporization*

This flux density is fairly high, because we are dealing with a phenomenon of boiling on contact with a hot wall.

For the type of products treated, which are of pasty consistency, [VAS 83] put forward an empirical expression for the heat flux density. The parameters of this equation vary depending on the nature of the product at hand.

$$\varnothing_E = \left[C_s^{0.5}, (T_s - T_e)^3, X \right] f(X)$$

$\varnothing_e$: heat flux density: $W.m^{-2}$;

T_s: temperature of the metal surface: °C;

T_e: boiling point of the product at working pressure: °C;

X: humidity over dry of the product to be treated;

C_s: surface charge of dry product: $kg.m^{-2}$

$$C_s = e\rho_s$$

e: thickness of the dried film: m;

ρ_s: density of the product in the dry state: $kg.m^{-3}$.

Heating vapor is used up to 3.5 bar relative, which is to say 145°C.

EXAMPLE 5.5.–

Let us look at the case of the aqueous dispersion of yeast:

$$\varnothing_E = \left[5000 + 0.95 (T_s - 100)^3 \right] \frac{(\sin(0.39X))^{2.2}}{C_s^{0.5}}$$

with:

$$X = 0.5 \qquad \rho_s = 1200 \text{ kg.m}^{-3}$$

$$T_s = 120°C \qquad e = 2 \times 10^{-4} \text{ m}$$

$$C_s = 1200 \times 2 \times 10^{-4} = 0.24 \text{ kg.m}^{-2}$$

$$\varnothing_E = \left[5000 + 0.95 \times 20^3 \right] \frac{\left[\sin \left(0.39 \times 0.5 \right) \right]^{2.2}}{0.24^{0.5}}$$

$$\varnothing_E = 703 \text{ W.m}^{-2}$$

5.6.3. Simulation

If R is the radius of the cylinder and ω its angular velocity of rotation, the elapsed time is:

$$\tau = \frac{x}{R\omega}$$

Here, x is the curvilinear abscissa expressing the progression of the product.

In the thickness of the metal wall, heat conduction is non-steady because the intensity of boiling varies with the humidity of the product, and that humidity is decreasing.

Fourier's law is written:

$$\frac{\partial T}{\partial \tau} = \frac{-\lambda}{C\rho} \frac{\partial^2 T}{\partial y^2}$$

y is the radial coordinate.

On contact with the product, the heat flux density is:

$$\varnothing_E = -\lambda \left[\frac{\partial T}{\partial y} \right]_{surface}$$

On the side of condensation of the heating vapor, the heat transfer coefficient ranges from 4,000 to 7,000 $W.m.^{-2}°C^{-1}$ depending on the concentration of inert species contained – i.e. depending on the setting of the purge valve.

In light of the low thickness of the product, it is possible to discount the variation in its sensible heat. The calculation is performed with no difficulty using the finite-difference method.

5.7. Fluidized continuous-flow dryer

5.7.1. *Principle*

This type of device is used to dry crystals which are fluidized in a thin layer on a rectangular table. The product progresses in a direction parallel to the longer side of the table. It is fed in along one of the shorter sides and overflows onto a spillway at the other end of the table. The spillway maintains the level of the fluidized bed.

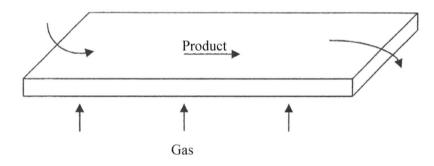

Figure 5.5. *Fluidization drying table*

The depth of the bed ranges from 5 to 30 cm, its length from 5 to 15 m, and its width from 1 to 2 m. The air injection grid is designed with a perforated surface fraction of around 12%.

The size of the particles treated by a fluidized bed ranges from 0.1 to 3 mm, or sometimes even larger. The velocity of the air is between 1.5 and

2 times the velocity at the beginning of fluidization. The length of stay of the crystals varies between 5 and 30 minutes depending on the nature of the product under treatment.

Crystals are products such that bubbles of gas tend to form in the bed as soon as fluidization begins. These bubbles correspond to a flow of gas which passes through the bed without producing any effect. To lessen the extent of this phenomenon, one subjects the table to a vibration movement, characterized by the ratio $a\omega^2/g$, where a is the amplitude of vibration and ω the pulsation. It is pointless for $a\omega^2/g$ to be any greater than 5, but necessary that it be greater than 1. The pulsation must be no greater than 25 Hz and the amplitude a is of the order of a few cm.

For example:

$$\frac{a\omega^2}{g} = \frac{0.03 \times 20^2}{9.81} \#1.2$$

In addition, the vibration of the fluidized bed disrupts the accumulation of sticky or cohesive crystals, as wet crystals may be, and prevents them from adhering to the walls.

[REA 78] gives an example of the calculations for such a dryer.

5.8. Drying of porous substances

5.8.1. *Definitions*

We distinguish between:

1) Shrinkable products: these products are very often foodstuffs. As the liquid leaves the porous substance, that substance's texture becomes denser, so its porosity decreases and no gas is able to penetrate through to the inside of the solid.

2) Ceramic-type products. During drying (pre-drying before high-temperature kilning), these products practically do not shrink at all, and it is the liquid surface (we speak of the liquid "front") which gradually retreats

from the surface. The liquid front is irregular and includes fingers of gaseous phase pushing into the liquid phase much like a glove, but the length of those "fingers" is no more than ten times the diameter of a particle. Therefore, we do not need to take it into account.

3) Density of an ideal gas

$$\rho_G = \frac{M_G P}{RT} \qquad \left(kg.m^{-3} \right)$$

M_G: molar mass of the gas: $kg.kmol^{-1}$;

R: ideal gas constant: $8314 \ J.kmol^{-1}.K^{-1}$;

T: absolute temperature;

P: pressure of the gas: Pa.

4) Heat conductivity of a porous substance. This property can be approximated by the following formula:

$$\lambda = \left[\lambda_G^n (1-S)\varepsilon + \lambda_L^n S\varepsilon + \lambda_S^n (1-\varepsilon) \right]^{1/n} \qquad \left(W.m^{-1}{}^\circ C^{-1} \right)$$

where:

λ_G, λ_L, λ_S: conductivity of the gas, of the liquid and of the solid;

ε: porosity of the porous substance;

S: saturation of the non-solid space: dimensionless.

5) Arrangement of the porous solid

The porous solid, for our purpose, will take the form of a flat plate divided into n slices of equal thickness, with the current indicator i. The hot air (or gas) sweeps the free surface of the plate. The other face is thermally insulated by a wall impermeable to the liquid and to gases.

We assign the index i to the properties of the fluids *at the center* of the segment whose index is i. The total n may be of the order of 10 or 15.

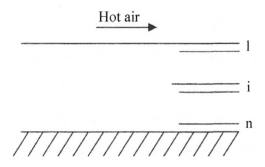

Figure 5.6. *Plate of porous solid subjected to drying by sweeping with hot air*

5.8.2. *Drying of shrinkable products*

There is no gaseous phase inside the plate subject to drying. Vaporization takes place entirely at the free surface of the solid.

The mean pore radius is [DUR 99]:

$$R = \frac{\varepsilon d_p}{3(1-\varepsilon)}$$

The porosity ε, for our purpose, corresponds to the volume not occupied by the solid.

The diameter d_p is imaginary and is of the order of magnitude of the size of a cell of living tissue (between 1 and 10 μm).

We note that if the porosity decreases, R does likewise, whereas the capillary pressure P_C increases.

$$P_C = \frac{\varepsilon \gamma \cos\theta}{R} = \frac{6\gamma \cos\theta (1-\varepsilon)}{\varepsilon d_p}$$

θ: angle of wetting of the solid by the liquid: rad;

γ: surface tension of the liquid: N.m^{-1}.

At the surface, the flux density of vaporization is (see section 2.2.8 in [DUR 16c]):

$$W_V = \beta \frac{M}{RT} \left(a_E \pi(t_S) - P_C - P_{VG} \right) \qquad \left(kg.m^{-2}.s^{-1} \right)$$

The activity a_E of the water can be measured using the proximity equilibrium cell method developed by [LEN 83], and modified by [PAL 87]:

t_S: surface temperature of the solid: °C;

$\pi(t_S)$: saturating vapor pressure of the water at t_S: Pa;

P_{VG}: partial pressure of water vapor in the sweeping air: Pa;

β: material transfer coefficient (see sections 2.3.1 and 2.3.3 in [DUR 16]): m.s^{-1}

The heat necessary for evaporation is transmitted by convection, and we have:

$$W_V = \frac{\alpha}{r}\left(t_G - t_S \right)$$

r: latent heat of vaporization: J.kg^{-1};

α: heat transfer coefficient: (see section 2.3.1 in [DUR 16]): W.m^{-2}.°C^{-1}

t_G: temperature of the gas: °C.

By writing that these two values of W_V are equal, we obtain the equation giving the surface temperature t_S.

The variation of ε over the time period $\Delta\tau$ is:

$$h\Delta\varepsilon = \frac{W_V}{\rho_L}\Delta\tau$$

We have been working on the (rather inexact) assumption that the decrease in porosity (and therefore in humidity) was constant over the thickness of the plate.

Let us stress the fact that as time passes, the depression P_C increases, the activity of the water decreases and, consequently, the evaporation rate W_V also decreases.

5.8.3. *Critical humidity*

In zone 1, which is the precritical drying zone, evaporation only takes place on the surface, and the surface is completely saturated with water. The surface temperature is the humid temperature of the air (see section 4.2.9).

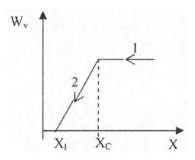

Figure 5.7. *Critical humidity X_C*

In the postcritical zone 2, the evaporation rate falls at the same time as the humidity X. It is possible to take account of this simply by conserving the law of surface evaporation surface (see section 5.8.1) and accepting that the coefficient β decreases with X in a linear fashion.

$$\beta_2 = \beta\left(\frac{X - X_1}{X_C - X_i}\right)$$

X_C: critical humidity;

X_i: any irreducible humidity.

Nadeau *et al.* [NAD 95], in their book, put forward a series of theoretical and empirical relations, particularly for:

– infrared-dried paper or film;

– percolation of hot air through a fixed bed;

– air-sweeping of a layer of sand;

– drying of wood.

5.8.4. *The solid is surrounded by a skin*

A solid such as this might be:

– a sausage;

– a fruit (grape, apricot, prune, fig, etc.).

The skin acts to provide addition resistance to:

– material transfer R_m;

– heat transfer R_t.

This is tantamount to replacing β with B where:

$$\frac{1}{B} = \frac{1}{\beta} + R_m$$

and replacing α with A where:

$$\frac{1}{A} = \frac{1}{\alpha} + R_t$$

We then need to hope that R_m and R_t depend on neither the humidity nor the temperature of the product.

5.8.5. *Ceramics–simplified method*

1) The heat necessary to vaporize the water is brought to the liquid front by conduction from the surface across the dry part. We have:

t_S: surface temperature of the dry solid: °C;

λ: heat conductivity of the dry part (see section 4.8.1): $W.m^{-1}.°C^{-1}$;

h_i: thickness of the dry part between the surface of the solid and the segment whose index is i;

t_{Fi}: temperature of the liquid front: °C.

The heat flux density transmitted by conduction is:

$$q = \frac{\lambda(t_S - t_F)}{h_i} \qquad \left(W.m^{-2}\right)$$

However, we also have:

$$q = \alpha(t_G - t_S)$$

We can write:

$$q = \frac{t_G - t_S}{\dfrac{1}{\alpha}} = \frac{t_S - t_F}{\dfrac{h_i}{\alpha}} = \frac{t_G - t_F}{\dfrac{1}{\alpha} + \dfrac{h_i}{\lambda}}$$

2) The vaporization flow of the liquid at the level i must cross a dry thickness equal to h_i.

We could write the vapor flux density expressed per unit surface of the solid:

$$W_V = \frac{\varepsilon D}{th_i} \frac{M}{RT}\left(\pi(t_F) - P_{VS}\right) \qquad \left(kg.m^{-2}.s^{-1}\right)$$

D: diffusivity of the vapor in the sweeping gas: $m^2.s^{-1}$;

ε: porosity of the solid: dimensionless;

t: here, the tortuosity of the solid: dimensionless

t#1.5

h_i: distance from the surface to the middle of the segment with index i: m;

$\pi(t_F)$: saturating vapor pressure of the liquid at t_F;

t_F: temperature of the front;

P_{VS}: partial vapor pressure at the surface of the solid: Pa.

A second expression for W_V is:

$$W_V = \beta \frac{M}{RT}(P_{VS} - P_{VG})$$

P_{VG}: partial vapor pressure in the sweeping gas: Pa

We can write:

$$W_V = \frac{P_{VS} - P_{VG}}{\dfrac{RT}{\beta M}} = \frac{\pi(t_F) - P_{VS}}{\dfrac{th\ RT}{\varepsilon DM}} = \frac{\pi(t_F) - P_{VG}}{\dfrac{RT}{M}\left(\dfrac{1}{\beta} + \dfrac{th}{\varepsilon D}\right)}$$

Knowing that $W_V = q/r$.

r: latent heat of vaporization of the liquid: J.kg^{-1}.

We obtain the equation which can used to calculate t_F.

$$W_V = \frac{t_G - t_G}{r\left(\dfrac{1}{\alpha} + \dfrac{h_i}{\lambda}\right)} = \frac{\pi(t_F) - P_{VG}}{\dfrac{RT}{M}\left(\dfrac{1}{\beta} + \dfrac{th}{\varepsilon D}\right)}$$

We can see that, as drying continues, the depth h_i increases, which decreases the vaporization rate W_V.

NOTE.–

The rigorous way to treat the drying of a porous substance is indicated in section 4.8.3.

5.9. Drying in a tunnel kiln

5.9.1. *Practical disposition*

The product in the state of distinct objects may be arranged on plates stacked on chariots or wagons.

If we are dealing with a divided solid (generally granulated), it may be arranged on a conveyor belt or on plates. The belt or the bottom of the plates are perforated and the hot gas circulates from top to bottom (to prevent the dust from flying away).

In tunnel dryers, a precise program of the temperature and humidity of the gas must be respected if we wish to prevent cracks (such is the case, for example, with the drying of certain ceramics).

For this type of reason, the circulation of the gas can be designed throughout the length co-current with the product or indeed counter-current or co-current in the first section and countercurrent thereafter.

In addition, to save on energy, we can recycle a portion of the partially-cooled air exiting the kiln instead of using outside air, which would be colder and therefore would require greater power to heat it. Remember that, to evaluate the characteristics of mixtures of air, the humid air diagram (Mollier diagram) is useful because, on that diagram, the image of a mixture is the barycenter of the images of the components assigned their respective masses.

Industrial realizations of dryers and tunnel kilns are both numerous and varied.

EXAMPLE 5.6.–

Data for the problem regarding a tunnel kiln.

We shall draw inspiration from the case described by [DAS 69]. It is a device designed to dry bricks.

The section of the tunnel is $0.87 \times 1.62 = 1.41$ m^2. The length of stay of the bricks in that kiln is 12 hours. They are laid on frames each containing 6 stages. Each stage contains 7 bricks laid on their short side, with the long side being directed along the length of the dryer.

The dimensions of a brick are $240 \times 115 \times 63$ mm^3. The mass of a brick at the output is 3.28 kg. The dryer has a capacity of 500 bricks per hour. The heating fluid is composed of fumes obtained from the combustion of natural gas. The air is taken in at 20°C and 50% relative humidity.

The fumes at the flowrate of 2.20 kg.s^{-1} result from the combustion of natural gas with 49% excess air. A heat balance which has nothing to do with drying shows that the temperature of the fumes is then 150°C.

These fumes exit the dryer at 60°C. The bricks, for their part, are heated from 50°C to 120°C.

The humidity of the air used for the combustion, read from a Mollier diagram, is 0.008.

The humidity of the bricks is 0.2 at the inlet and 0.05 at the outlet. The gravimetric specific heat capacity of the dry product is 920 J.kg^{-1}.°C^{-1}.

Our aim is to calculate:

– the characteristics of the fumes;

– the convection coefficient.

We then need to find (using the equivalent-exchanger method):

– the flowrate of water evaporated;

– the flowrate of fumes needed.

1) Characteristic values of the fumes

The specific flowrate of fumes per kg of natural gas is obtained by considering the natural gas to be pure methane:

$$g_F = 16.2$$

The mass of water vapor resulting from the combustion is:

$$g_V = 9 \times h = 9 \times 0.25 = 2.25$$

The stoichiometric air is:

$$a_S = 2.25 + 16.2 - 1 = 17.45$$

The excess air is:

$$a_e = 0.49 \times 17.45 = 8.55$$

The humidity of the fumes is therefore:

$$Y_F = \frac{2.25 + (17.45 + 8.55)\,0.008}{8.55 + 16.2}$$

Thus:

$$Y_F = 0.10$$

2) Convection coefficient

The area of the bricks perpendicular to the kiln is:

$$6 \times 7 \times 0.115 \times 0.063 = 0.3\,\text{m}^2$$

The section free for the passage of the air is therefore:

$$1.41 - 0.30 = 1.11\,\text{m}^2$$

The mass flux of fumes passing over the bricks is:

$$G = 2.20 / 1.11 = 1.98\,\text{kg.s}^{-1}.\text{m}^{-2}$$

Hence the coefficient:

$$\alpha = 14.4(1.98)^{0.8} = 24.87\,\text{W.m}^{-2}.°\text{C}^{-1}$$

3) Flowrate of water evaporated

The kiln contains $500 \times 12 = 6000$ bricks.

The active surface of each brick is:

$$2\times0.115\times0.24+0.063\times0.24=0.07\ \text{m}^2$$

The transfer surface area is:

$$S=0.07\times6000=420\text{m}^2$$

The log mean temperature difference between the fumes and the bricks is:

$$\Delta T=18.2°C$$

The flowrate of product is:

$$P=\frac{500}{3600}\times\frac{3.280}{1+0.05}=0.433\ \text{kg.s}^{-1}$$

Finally:

$$\Delta E=\frac{24.87\times420\times18.2-0.433\big(920+0.05\times4180\big)\big(120-50\big)}{2.5\times10^6+1826\times60-4180\times50}$$

$$\Delta E=0.065\ \text{kg.s}^{-1}$$

The final residual humidity is:

$$0.20-\frac{0.065}{0.433}=0.05$$

So, we find the experimental value which was given.

4) Flowrate of fumes necessary

The heat ceded by the fumes, per kg of dry fumes, is:

$$\big(150-60\big)\times\big(1000-1880\times0.11\big)=106,920\ \text{J.kg}^{-1}$$

The heat lost by the fumes through convection is:

$$\alpha S\Delta T=24.87\times420\times18.2=190,106\ \text{W}$$

The flowrate of dry fumes is therefore:

$$W_{FS} = \frac{190,106}{106,920} = 1.78 \text{ kg.s}^{-1}$$

The flowrate of humid fumes is:

$$W_F = W_{FS} \times \left[\frac{16.2 + 2.25 + 0.008(17.45 + 8.55)}{16.2} \right]$$

$$W_F = 2.05 \text{ kg.s}^{-1}$$

The error with respect to the true flowrate is:

$$\frac{2.20 - 2.05}{2.20} = 7\%$$

We can put this error down to the heat losses.

5.10. Heat transfer in drum dryers

5.10.1. *Surface of contact between product and gas*

Experience proves that contact between the gas and the solid particles is encouraged if lifting blades are installed in the dryers. The number N of those blades ranges between 5D and 10D, where D is the diameter of the drum expressed in meters. When the wall rotates, the product falls from the blades, forming a curtain of particles. The kiln is heat-proofed for the safety of the operators.

In order for the fall of the solid particles to be prolonged over the greatest possible portion of the upper half-circumference, it is advantageous to use bent blades. However, if the product is even a little sticky, only flat blades can be envisaged.

Lifting blades do have disadvantages: maintenance operations are more complicated, the manufacture price is higher and they exacerbate the problem of breaking of the particles by shock.

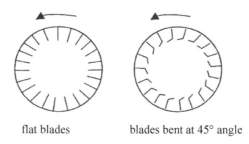

flat blades   blades bent at 45° angle

Figure 5.8. *Lifting blades (inclination)*

From the point of view of heat transfer, the improvement obtained is limited, because the true coefficient is much smaller than that of an isolated particle in a gas stream. Thus, we simply agree that the gas flows over both faces of each curtain of particles, without greatly dispersing the particles as it does so.

The number of curtains is, more or less, equal to half the number of layers N_a, and the number of faces is double the number of curtains – i.e. N_a in total. By comparing the surface of a rectangle with that of a circle, we can see that the mean height of the curtains is $\pi D/4$, where D is the diameter of the drum. Thus, the surface of contact with the gas is, per meter length of the drum, $N_a \pi D/4$, and to this we must add the free surface of the product, which is a fraction F_L of that of the drum, so we have $F_L \pi D$.

In total, the surface available for contact is:

$$S = \left[F_L + \frac{N_a}{4} \right] \pi DL$$

5.10.2. *Heat transfer coefficient*

When a gas flows across a flat or slightly-curved surface, the convection coefficient is:

$$\alpha = 14.4 \phi_G^{0.8} \qquad \left(W.m^{-2}.C^{-1} \right)$$

ϕ_G: mass flux density of the gas: $kg.m^{-2}.s^{-1}$.

This expression is useful, *a priori*, for heat exchange between, firstly, the gas and, secondly, either the wall of the drum or the free surface of the product.

In the particular case of a curtain of flowing particles, we can accept that the gas stream causes a moderate degree of disturbance to the trajectory of the outermost particles, which is likely to increase the overall coefficient by around 50%.

In addition, according to some, the exponent of ϕ_G for rotary drums is 0.67 rather than 0.8. If we accept a mean value of $1\text{kg.m}^{-2}.\text{s}^{-1}$ for ϕ_G and, in light of the aforementioned 50% increase, the above formula becomes:

$$\alpha = 22 \ \phi_G^{0.67}$$

NOTE.–

[KRA 52] recap Vahl's method for the displacement of the solid and deduce the retention in an inclined rotary kiln.

5.10.3. *Example of simplified thermal calculation of a bladed dryer*

We shall use the equivalent-exchanger method.

Suppose that the air available is at 190°C, that the product enters the dryer at 100°C and that drying is countercurrent.

It is conventional to accept that the product exits at a temperature 10–20°C below that of the air. In our case, the product will exit at 170°C. At the other end, we shall accept that the air exits at a temperature 50°C higher than that of the product, meaning that it exits at 150°C.

The temperature program is summarized as follows:

Air	190	$\longrightarrow$	150
Crystals	170	$\longleftarrow$	100
	20		50

The LMTD (log mean temperature difference) is 32.71°C.

Let us now calculate the thermal power necessary for drying.

The flowrate of solid is 150 kg.h^{-1} and the humidity to be evacuated represents 2% of the solid. To evaporate water at 100°C and superheat it to 120°C, we must apply 2.3×10^6 J.kg^{-1}.

Hence:

$$\frac{150}{3600} \times 0.02 \times 2.3 \times 10^6 = 1916 \text{ W}$$

The sensible heat of the solid is calculated in the knowledge that its specific heat capacity is 600 J.kg^{-1}.°C^{-1}.

$$\frac{150}{3600} \times 600 \times (170 - 100) = 1750 \text{ W}$$

The heat losses from a non-insulated dryer can be obtained using a calculation of free convection which gives us a coefficient of 7 W.m^{-2}.°C^{-1}. The coefficient drops to 0.7 W.m^{-2}.°C^{-1} if the kiln is coated with glass wool.

Suppose that the device is an insulated rotary kiln, 0.7 m in diameter and 6 m long. The LMTD between the air and the surrounding environment at 20°C is equal to 149°C. The losses are therefore:

$$\pi \times 0.7 \times 6 \times 0.7 \times 149 = 1376 \text{ W}$$

The total heat required is:

$$1916 + 1750 + 1376 = 5042 \text{ W}$$

The flowrate of air is:

$$G = \frac{5042}{1000 \times (190 - 150)} = 0.124 \text{ kg.s}^{-1}$$

The number of fins will be:

$0.7 \times 10 = 7$, so an even number of 8 fins

The product αS of the transfer coefficient by the surface area of contact can be obtained by hypothesizing that the free surface of the product is 20% of the side surface of the drum:

$$\alpha S = 22 \left(0.2497\right)^{0.67} \left[0.2 + \frac{8}{4} \right] \times \pi \times 0.7 \times 6$$

$$\alpha S = 15943 \ W.°C^{-1}$$

Hence, we have the available power:

$$\alpha S \Delta T = 159.43 \times 32.71 = 5220 \ W$$

This power exceeds the value of 4885 W which we need, and the safety margin is:

$$\frac{5220 - 5042}{5042} = 3.5\%$$

In light of the poor thermal power transmitted relative to the financial investment required, these dryers are not widely used.

5.11. Ancillary installations

5.11.1. *Dust removal from air*

This operation can be carried out using one or more of the following devices:

– cyclone;

– wet washer;

– sleeve filter;

– electrofilter.

The cyclone generally needs to be used if the dust content is any higher than 0.2 kg.m^{-3}. In general, at the outlet from a cyclone, the load in terms of fine particles is no greater than 130–200 mg.m^{-3}.

The detailed work is carried out by one of the three other devices. A sleeve filter or an electrofilter stop practically all the fine particles. A wet washer allows 20–30 mg.m^{-3} of dust to pass through.

The sleeves of the filter must be cleared of the dust that they have stopped when the load loss through the filter reaches 1500 Pa [DUR 99].

5.11.2. *Cooling of the product*

Too high a temperature of the outgoing product may prevent its bagging (in plastic sacks) or may cause a slow deterioration of the product during storage. It is therefore necessary to cool the product. Let us give a few examples.

1) At the outlet of a fluidized table

Cooling may be done by sliding over an inclined, double-wall plane, cooled by water. However, it is necessary that the product should not cluster under the influence of the condensation of the ambient water vapor on the inclined plane.

Following the output from the dryer, we could also put in place a second fluidized table, fed with cold air.

2) After a pneumatic dryer

A pneumatic transport with cold air is apt, particularly if it finishes with water-cooled double-wall cyclones.

3) After a rotary drum

We can use a second drum fed with cold air.

4) After an atomization tower

It is possible to use a fluidized bed on a table or in a vat, but the most usual solution is to install a pneumatic transport which simultaneously resolves the problem of transport in a silo.

In any case, the cooling air must have sufficiently low humidity to prevent the agglomeration of the crystals. For certain hygroscopic products, we pre-treat the air so that its dew point is -40°C.

In conclusion, note that the equivalent-exchanger approximation is rigorous for the calculations for a cooler.

5.12. Choice of dryers

5.12.1. *Depending on the physical state of the product*

We distinguish:

1) Pasty or syrupy products which are treated, depending on the case, with:

– a cylinder dryer;

– a tunnel dryer with a conveyor belt where the heat is applied either by radiation or by gaseous convection, or indeed by conduction across the belt, which is then made of metal;

– an atomization tower;

– a vat (often in a vacuum) for batch operating;

– a cylinder with a heated wall, internally stirred by inclined pallets or a screw. These devices are generally used only for pre=drying.

2) Divided solids are treated:

– in a continuous-flow fluidized bed;

– in a pneumatic dryer;

– sometimes in a direct-heating rotary drum;

– in an indirect-heating rotary drum if the product is rich in fine particles;

– by radiation (infrared or microwave) in discontinuous vats or continuous tunnels;

– by gaseous convection also in batch vats or continuous tunnels.

Microwave drying is unsuitable for divided solids, because the porosity of the bed (around 0.4–0.5) decreases the conductivity and hampers thermal homogenization.

3) Compact solids. This term applies to, say, pharmaceutical tablets, pieces of ceramic or indeed pieces of wood. The drying takes place in:

– continuous-flow tunnels operating by radiation or gas convection. The baskets, plates or pallets are laid on chariots which progress slowly;

– discontinuous vats with immobile baskets or frames;

– in open air or in vats for wood.

NOTES.–

Firing kilns for ceramics, and particularly for porcelain at 900°C or 1,300°C ("furnace") are either electric ovens or gas-fired muffle kilns (with a radiating wall).

Cement kilns are fairly similar to calcination kilns, as they are rotary kilns where the clinker is in direct contact with combustion fumes. We shall discuss calcination kilns in Chapter 1 in [DUR 16e].

However, these can no longer be categorized as drying devices.

5.12.2. *Depending on the energy expenditure*

Gas-swept dryers consume between 3.5 and 12 MJ per kg of water evaporated, whereas conduction dryers (cylinders or belts) and radiation dryers need only 2.5–6 MJ per kg. In particular, the yield of microwave drying is excellent (for very wet and compact products).

5.12.3. *Depending on the intended fate of the product in the dryer*

1) Breaking of crystals.

Breakage is easier to achieve when the crystals have poor shock resistance.

The fluidized dryer, where the lengths of stay are long (up to 1 hour), rounds and polishes the crystals but does not break them. It produces fine particles.

A pneumatic dryer breaks the crystals (introduction with a possible clod crusher, bends, cyclone) and does not smooth them. The mean size of the crystals may be reduced by up to 40% in the worst cases.

The bladed rotary drum is intermediary between the former two solutions.

Vats and radiation dryers cause no mechanical damage to the crystals.

2) PSD of the feed.

A fluidized dryer cannot be used for crystals smaller than 100 μm, because the entrainment would be excessive.

A pneumatic dryer can deal with all PSDs, because it can easily entrain the fine particles.

In the rotary drum, the air velocity is limited by the entrainment of the fine particles. This increases the diameter of the device, and consequently the investment needed. The same is true of a tunnel dryer. Thus, an indirect-contact rotary dryer may present the best solution.

3) Temperature sensitivity.

The pneumatic dryer is ideal because it is co-current and the length of stay is approximately one second.

The rotary drum may work co-current, but the length of stay is tens of minutes.

The fluidized bed is not recommended, unless we use air at a sufficiently-moderate temperature, but in that case, the flowrate of air is high.

The cylinder dryer is well suited for thermosensitive products.

5.12.4. *Depending on the hygroscopicity of the product*

This criterion pertains only to crystals. If the activity of the water in the parent liquor is much less than 1, the product will be difficult to dry. We then say that it is hygroscopic.

The length of stay in a pneumatic dryer is too short for a product that is difficult to dry. On the other hand, the batch fluidized bed and the continuous-flow rotary drum may be appropriate, because the respective lengths of stay may be up to 30 minutes or 1 hour.

5.12.5. *Depending on the flowrate treated*

Certain continuous-flow devices are incapable of processing more than 5 tons per hour. They are:

– the pneumatic dryer;

– the fluidized bed;

– the cylinder dryer;

– the tunnel dryer.

Certain atomization towers can treat up to 15 tons.h^{-1}, and the production of rotary drums may reach considerable values (cement kilns).

5.12.6. *Depending on the price*

Charge dryers ("discontinuous flow") and pneumatic dryers are the least expensive.

5.12.7. *Depending on the inlet temperature of the gas*

With very fine products (such as activated charcoal), we can use dryers with superposed circular plates, raked by rotating blades. The product flows, under the influence of gravity, from one plate to the one immediately beneath it. This type of dryer is expensive to buy but can withstand very high temperatures.

Rotary dryers and superposed-plate dryers are also capable of using high air temperatures.

5.13. A few energy considerations

5.13.1. *Thermal recovery by a heat pump*

The hot, humid air from the dryer evaporates a refrigerant liquid in the exchanger E_1 and, in doing so, a portion of the humidity it carries condenses and is extracted.

This humidity represents the evaporation in the dryer.

The refrigerant gas is compressed by the compressor P, so that it condenses in E_2. For its part, the recycled air is heated up in E_2.

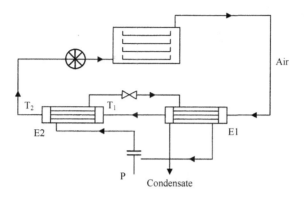

Figure 5.9. *Heat pump*

For a reversible Carnot cycle working between temperatures T_1 and T_2, the ideal productivity is:

$$\frac{T_2}{T_2 - T_1}$$ (section 1.7.1 vol. 1)

Thus, if $T_2 = 350$ K, $T_1 = 330$ K, the productivity is:

$$\frac{350}{350 - 330} = 17.5$$

If 1 kW is consumed by the compressor P, we should obtain 17.5 kW of heat in the air. In practice, though, the productivity is three times less.

The proliferation of this procedure is hampered by the size of the compressors that are needed. In each case, an economic balance needs to be struck. Finally, note that the level of humidity in the air loop is much higher than it is with open-circuit operation, which does not lend itself well to drying. Otherwise, we need to significantly lower the temperature of E_1, which is not economical.

5.13.2. *Drying with solvent-recovery*

When the solvent evaporated during the drying process is not water, but instead an expensive product, it is normal to try to recover it at the end of the process. In these conditions, traditional drying in an air stream leads to anti-economical recovery, because it will involve:

– either condensation in the presence of a considerable mass of inert species;

– or absorption in a fairly automated (and therefore expensive) installation.

In reality, it is preferable to carry out drying with a vapor that is likely to condense.

Generally, solvents are organic products which are not highly soluble in water. Therefore, it is enough to pass over the product with superheated water vapor whose sensible heat is great enough to evaporate the liquid solvent. At output, we then collect a gaseous mixture of solvent vapor and water vapor, which condenses easily. The solvent is then recovered by decantation. Furthermore, if the procedure is designed, from the get-go, with a non-toxic, non-flammable solvent, the installation will be greatly simplified.

APPENDICES

Appendix 1

Numerical Integration
4th-order Runge–Kutta Method

The aim here is to integrate the differential equation:

$$\frac{dx}{d\tau} = F(x, \tau)$$

$$x_{\tau=0} = x_0$$

We set:

$$\tau_{i+1/2} = \tau_i + \frac{\Delta\tau}{2}$$

$$x_{i+1/2}^{(1)} = x_i + \frac{\Delta\tau}{2} F(x_i, \tau_i)$$

$$x_{i+1/2}^{(2)} = x_i + \frac{\Delta\tau}{2} F(x_{i+1/2}^{(1)}, \tau_{i+1/2})$$

$$x_{i+1}^{(1)} = x_i + \Delta\tau\, F(x_{i+1/2}^{(2)}, \tau_{i+1/2})$$

Therefore:

$$x_{i+1} = x_i + \frac{\Delta\tau}{6}\Big[F(x_i, \tau_i) + 2F(x_{i+1/2}^{(1)}, \tau_{i+1/2})$$

$$+ 2F(x_{i+1/2}^{(2)}, \tau_{i+1/2}) + F(x_{i+1}^{(1)}, \tau_{i+1})\Big]$$

The method can be generalized to apply to a system of n 1st-order differential equations involving n variables x_j (j ranging from 1 to n). The independent variable is x_0.

$$\frac{dx_j}{dx_0} = F_j\left(x_0, x_1, ..., x_j, ..., x_n\right)$$

Let us set:

$$x_{0,i+1} = x_{0,i} + \Delta x_0 \quad \text{and} \quad x_{0,i+1/2} = x_{0,i} + \frac{\Delta x_0}{2}$$

$$x_{j,i+1/2}^{(1)} = x_{j,i} + \frac{\Delta x_0}{2} F_j\left(x_{0,1}, ..., x_{j,i}, ... x_{n,i}\right)$$

$$x_{j,i+1/2}^{(2)} = x_{j,i} + \frac{\Delta x_0}{2} F_j\left(x_{0,i+1/2}, ..., x_{j,i+1/2}^{(1)}, ..., x_{n,i+1/2}^{(1)}\right)$$

$$x_{j,i+1}^{(1)} = x_{j,i} + \Delta x_0 F\left(x_{0,i+1/2}, ..., x_{j,i+1/2}^{(2)}, ..., x_{n,i+1/2}^{(2)}\right)$$

and finally:

$$x_{j,i+1} = x_{j,i} + \frac{\Delta x_0}{6}\left[F_j\left(x_{0,i}, ... x_{j,i}, ..., x_{n,i}\right) + 2F_j\left(x_{0,i+1/2}, ... x_{j,1+1/2}^{(1)}, ... x_{n,i+1/2}^{(1)}\right) \right.$$
$$\left. + 2F_j\left(x_{0,i+1/2}, ... x_{j,i+1/2}^{(2)}, ... x_{n,i+1/2}^{(2)}\right) + F_j\left(x_{0,i+1}, ... x_{j,i+1}^{(1)}, ... x_{n,i+1}^{(1)}\right) \right]$$

Appendix 2

A Few Important Identities

Theorem of reciprocity

We suppose that:

$$F(x, y) = \text{const.} = C \quad \text{and} \quad dF = 0$$

x and y are not independent, and we can always write:

$$y = f(x) \quad \text{and} \quad dy = \left(\frac{d\,y}{dx}\right)_F dx$$

$$dF = \left(\frac{\partial F}{\partial x}\right)_y dx + \left(\frac{\partial F}{\partial}\right)_x dy = \left[\left(\frac{\partial F}{\partial x}\right)_y + \left(\frac{\partial F}{\partial}\right)_x \left(\frac{\partial}{\partial x}\right)_F\right] dx = 0$$

The expression in square brackets must be equal to zero, so:

$$\left(\frac{\partial F}{\partial x}\right)_y = -\left(\frac{\partial F}{\partial}\right)_x \left(\frac{\partial}{\partial x}\right)_F \qquad\qquad \text{[C.1]}$$

Similarly, were we to permute x and y, we would have:

$$\left(\frac{\partial F}{\partial}\right)_x = -\left(\frac{\partial F}{\partial x}\right)_y \left(\frac{\partial x}{\partial}\right)_F \qquad\qquad \text{[C.2]}$$

By multiplying, term by term, we obtain the theorem of reciprocity:

$$\left(\frac{\partial}{\partial x}\right)_F \left(\frac{\partial x}{\partial}\right)_F = 1 \quad \text{or indeed} \quad \left(\frac{\partial}{\partial x}\right)_F = \frac{1}{\left(\dfrac{\partial x}{\partial}\right)_F} \qquad [C.3]$$

Closed-loop derivation

In view of equations [C.1] and [C.3]:

$$\left(\frac{\partial F}{\partial}\right)_x \left(\frac{\partial}{\partial x}\right)_F \left(\frac{\partial x}{\partial F}\right)_y = -1$$

In a closed loop, the numerator of the first derivative is equal to the denominator of the third derivative.

Cascade derivation

We suppose that:

$$F = f(x) \quad \text{and} \quad x = g(u)$$

The cascade derivation is:

$$\frac{dF}{du} = \left(\frac{dF}{dx}\right)\left(\frac{dx}{du}\right)$$

This can be written as follows (theorem of reciprocity):

$$\left(\frac{dF}{dx}\right)_u \left(\frac{dx}{du}\right)_F \left(\frac{du}{dF}\right)_x = 1$$

NOTE.–

There is no contradiction between closed-loop derivation and cascade derivation. Indeed:

– in closed-loop derivation, one of the variables is constant and its total differential is zero;

– in cascade derivation, and for two out of the three derivatives, each variable is an explicit function of the next. The third derivative can be deduced from this.

We can see how important it is for the dependencies and constraints to be clearly defined.

Rational fractions and simple elements

Before performing any derivation calculation, and particularly before integration, it is *imperative* to break down the rational fractions into their simple elements. For example:

$$\frac{1}{g_2} = \frac{1}{(V - V_1)(V - V_2)} = \frac{1}{(V_1 - V_2)}\left[\frac{1}{(V - V_1)} - \frac{1}{(V - V_2)}\right]$$

$$\frac{1}{g_3} = \frac{1}{(V - V_1)(V - V_2)(V - V_3)} = \frac{1}{D_{12} \times D_{13}(V - V_1)} +$$

$$\frac{1}{D_{23} \times D_{21}(V - V_2)} + \frac{1}{D_{31} \times D_{32}(V - V_3)}$$

with:

$$D_{ij} = (V_i - V_j)$$

Bibliography

[ALÉ 90] ALÉON D., CHANRION P., NÉGRIÉ G. *et al.*, *Séchage du bois; Guide pratique*, Centre Technique du Bois et de l'Ameublement, Paris, 1990.

[ANT 89] ANTIA F.D., HORVATH C., "Gradient elution in non-linear preparative liquid chromatography", *Journal of Chromatography A*, vol. 484, pp. 1–27, 1989.

[BAD 66] BADDOUR R.F., U.S. Patent 3250058, 1966.

[BEN 86] BENNASRALLAH S., ARNAUD G., "Séchage d'un milieu poreux contenant une faible teneur en eau". *International Journal of Heat and Mass Transfer*, vol. 29, no. 9, pp. 1443–1449, 1986.

[BER 51] BERG C., "Hypersorption design. Modern advancements", *Chemical Engineering Progress*, vol. 47, p. 585, 1951.

[BOQ 78] BOQUET R., CHIRIFE J., IGLESIAS H.A., "Equations for fitting water sorption isotherms of foods. II. Evaluation of various two parameter models", *International Journal of Food Science & Technology*, vol. 13, p. 319, 1978.

[BOW 40] BOWMAN R.A., MUELLER A.C., NAGLE W.M., "Mean temperature difference in design", *Transactions of the American Society of Mechanical Engineers*, vol. 62, p. 283, 1940.

[BOY 47] BOYD G.E., ADAMSON A.W., MYERS L.S., "The exchange adsorption of ions from aqueous solutions by organic zeolites", *Journal of the American Chemical Society*, vol. 69, p. 2896, 1947.

[BRO 52] BROWNING F.M., "Vapour phase processes serve industry well", *Chemical Engineering*, vol. 59, pp. 158–161, 1952.

[BRO 64] BROEKHOFF J.C.P., DE BOER J.H., *Studies on pore systems in catalysts. Pore distribution calcul*, 1964.

[BRO 67a] BROEKHOFF J.C.P., DE BOER J.H., "Studies on pore systems in catalysts. Calculation of pore distribution. A. Fundamental equations", *Journal of Catalysis*, vol. 9, p. 8, 1967.

[BRO 67b] BROEKHOFF J.C.P., DE BOER J.H., "Studies on pore systems in catalysts. Calculation of pore distribution. B. Applications", *Journal of Catalysis*, vol. 9, p. 15, 1967.

[BRU 80] BRUIN S., LUYBEN K.CH.A.M., "Drying of food materials, a review of recent developments", in *Advances in Drying*, vol. 1, Hemisphere Publshing Corporation, Washington, DC, 1980.

[CAD 78] CADIERGUES R., "Propriétés de l'air humide et de l'eau, justification de nouveaux choix", *Promoclim E*, vol. 9E, no. 1, pp. 21–38, 1978.

[CAM 76] CAMPBELL J.C., "How to prevent cooling tower fog", *Hydrocarbon Processing*, pp. 97–100, 1976.

[CAR 50] CAREY W.F., WILLIAMSON G.J., "Gas cooling and humidification. Design of packed towers from small-scale tests", *Proceedings-Institution of Mechanical Engineers*, vol. 163, no. WEP 56, pp. 41–53, 1950.

[CER 70] CERRO R.L., SMITH J.M., "Chromatography of nonadsorbable gases", *AIChE Journal*, vol. 16, p. 1034, 1970.

[CHI 31] CHILTON T.H., COLBURN A.P., "Pressure drop in packed tubes", *Industrial and Engineering Chemistry*, vol. 23, no. 8, pp. 913–919, 1931.

[COL 31] COLBURN A.P., KING W.J., "Relationship between heat transfer and pressure drop", *Industrial and Engineering Chemistry*, vol. 23, no. 8, pp. 919–923, 1931.

[COL 67] COLLINS J.J., "The lub equilibrium section concept for fixed bed adsorption", *Chemical Engineering Progress Symposium Series*, vol. 63, p. 52, 1967.

[COO 65] COONEY D.O., Lightfoot E.N., "Existence of asymptotic solutions to fixed-bed separations and exchange equations", *Industrial & Engineering Chemistry Fundamentals*, vol. 4, p. 233, 1965.

[CRA 57] CRANSTON R.W., INKLEY F.A., "The determination of pore structures from nitrogen adsorption isotherms", *Advances in Catalysis*, vol. 9, p. 143, 1957.

[CYS 91] CYSERVSKI P., JAULMES A., LEMQUE R. *et al.*, "Multivalent ion exchange model of biopolymer chromatography for mass overload conditions", *Journal of Chromatography A*, vol. 548, p. 61, 1991.

[DAS 69] DASCALESCU A., *Le séchage et ses applications industrielles*, Dunod, 1969.

[DE 43] DE VAULT D., "The theory of chromatography", *Journal of the American Chemical Society*, vol. 65, p. 532, 1943.

[DE 68] DE MONBRUN J.R., "Factors to consider in selecting a cooling tower", *Chemical Engineering*, pp. 106–116, 1968.

[DE 80] DE JONG A.W.J., KRAAK J.C., POPPE N. *et al.*, "Isotherm linearity and sample capacity in liquid chromatography", *Journal of Chromatography A*, vol. 193, pp. 181–195, 1980.

[DUD 78] DUDUKOVIC M.P., LAMBA H.S., "Solution of moving boundary problems for gas–solid noncatalytic reactions by orthogonal collocation", *Chemical Engineering Science*, vol. 33, pp. 303–314, 1978.

[DUL 62] DULLIEN F.A.L., SCOTT D.S., "The flux ratio for binary counter diffusion of ideal gases", *Chemical Engineering Science*, vol. 17, pp. 771–775, 1962.

[DUL 70] DULLIEN F.A.L., BATRA V.K., "Determination of the structure of porous media", *Industrial & Engineering Chemistry*, vol. 62, no. 10, p. 25, 1970.

[DUR 99] DUROUDIER J.-P., *Pratique de la filtration*, Hermès, 1999.

[DUR 16a] DUROUDIER J.P., *Thermodynamics*, ISTE Press, London and Elsevier, Oxford, 2016.

[DUR 16b] DUROUDIER J.-P., *Distillation*, ISTE Press, London and Elsevier, Oxford, 2016.

[DUR 16c] DUROUDIER J.-P., *Liquid–Liquid and Solid–Liquid Extractors*, ISTE Press, London and Elsevier, Oxford, 2016.

[DUR 16d] DUROUDIER J.-P., *Liquid–Solid Separators*, ISTE Press, London and Elsevier, Oxford, 2016.

[DUR 16e] DUROUDIER J.-P., *Heat Transfer in the Chemical, Food and Pharmaceutical Industries*, ISTE Press, London and Elsevier, Oxford, 2016.

[ECK 80] ECKERT E.R.G., FAGHRI M., "A general analysis of moisture migration caused by temperature differences in an unsaturated porous medium", *International Journal of Heat and Mass Transfer*, vol. 23, p. 1613, 1980.

[ECK 86] ECKERT E.R.G., FAGHRI M., "A parametric analysis of moisture migration in an unsaturated porous slab caused by convective heat and mass transfer", *Wärme und Stoffübertragung*, vol. 20, pp. 77–87, 1986.

[EL 77] EL SABAAWI M., PEI D.C.T., "Moisture isotherms of hygroscopic porous solids", *Industrial & Engineering Chemistry Fundamentals*, vol. 16, no. 3, p. 321, 1977.

[ERM 61] ERMENC E.D., "Designing a fluidized adsorber", *Chemical Engineering*, vol. 29, p. 87, 1961.

[FAI 69] FAIR J.R., "Sorption processes for gas separation", *Chemical Engineering*, vol. 76, p. 90, 1969.

[FIN 65] FINLAYSON B.A., SCRIVEN L.E., "The method of weighed residuals and its relation to certain variational principles for the analysis of transport processes", *Chemical Engineering Science*, vol. 20, p. 395, 1965.

[FRÖ 38] FRÖSSLING N., "Ueber die Verdunstung fallender Tropfen", *Gerlands Beitrage zur Geophysik*, vol. 520, p. 170, 1938.

[FUR 68] FURZER I.A., "The natural draught cooling tower", *British Chemical Engineering*, vol. 13, p. 1298, 1968.

[GHR 88] GHRIST B.F.D., SNYDER L.R., "Design of optimized high-performance liquid chromatographic gradients for the separation of either small or large molecules", *Journal of Chromatography A*, vol. 459, p. 43, 1988.

[GIB 86] GIBBS S.J., LIGHTFOOT E.N., "Scaling up gradient elution chromatography", *Industrial & Engineering Chemistry Fundamentals*, vol. 25, p. 490, 1986.

[GLU 47] GLUECKAUF E., COATES J.I., "Theory of chromatography. Part 4. The influence of incomplete equilibrium on the front boundary of chromatograms and on the effectiveness of separation", *Journal of the Chemical Society*, pp. 1315–1321, 1947.

[GLU 49] GLUECKAUF E., "Theory of chromatography. Part 6 Precision measurements of adsorption and exchange isotherms from column elution data", *Journal of the Chemical Society*, pp. 3280–3285, 1949.

[GLU 50] GLUECKAUF E., "Theory of chromatography. Part. 9. The theoretical plate concept in column separation", *Transactions of the Faraday Society*, vol. 51, p. 34, 1950.

[GLU 55] GLUECKAUF E., "Theory of chromatography Part 10. Formulae for diffusion into spheres and their application to chromatography", *Trans. Faraday Soc.*, vol. 51, p. 1540, 1955.

[GOL 53] GOLDSTEIN S., "On the mathematics of exchange processes in fixed columns. I. Mathematical solutions and asymptotic expansions", *Proceedings of the Royal Society of London A*, vol. 219, p. 151, 1953.

[GOL 88] GOLSHAN-SHIRAZI S., GUIOCHON G., "Analytical solution for the mode of chromatography in the case of a Langmuir isotherm", *Analytical Chemistry*, vol. 60, p. 2364, 1988.

[GOL 89a] GOLSHAN-SHIRAZI S., GUIOCHON G., "Analytical solution for the ideal model of elution chromatography in the case of a binary mixture with competitive Langmuir isotherm", *Journal of Chromatography A*, vol. 484, p. 125, 1989.

[GOL 89b] GOLSHAN-SHIRAZI S., GUIOCHON G., "Analytical solution for the ideal model of chromatography in the case of a pulse of a binary mixture with competitive Langmuir isotherm", *Journal of Physical Chemistry*, vol. 93, p. 4143, 1989.

[GOL 89c] GOLSHAN-SHIRAZI S., GUIOCHON G., "Theory of optimisation of the experimental conditions of preparative elution using the ideal model of liquid chromatography", *Analytical Chemistry*, vol. 61, p. 1276, 1989.

[GOL 91] GOLSHAN-SHIRAZI S., EL FALLAH M.Z., GUIOCHON G., "Effect of the intersection of the individual isotherms in displacement chromatography", *Journal of Chromatography A*, vol. 541, p. 195, 1991.

[GOU 65] GOUDET G., *Les fonctions de Bessel et leurs applications en physique*, Masson, 1965.

[GRA 53] GRAHAM D., "The characterization of physical adsorption systems. I. The equilibrium function and standard free energy of adsorption", *Journal of Physical Chemistry*, vol. 57, p. 665, 1953.

[GRA 66] GRANT R.J., MANES M., "Adsorption of binary hydrocarbon gas mixtures on activated carbon", *Industrial & Engineering Chemistry Fundamentals*, vol. 5, p. 490, 1966.

[GUI 88] GUIOCHON G., GOLSHAN-SHIRAZI S., JAULMES A., "Computer simulation of the propagation of a large concentration band in liquid chromatography", *Analytical Chemistry*, vol. 60, pp. 1856–1866, 1988.

[GUN 80] GUNN D.J., "Theory of liquid phase dispersion in packed columns", *Chemical Engineering Science*, vol. 35, p. 2405, 1980.

[GUN 87] GUNN D.J., "Axial and radial dispersion in fixed beds", *Chemical Engineering Science*, vol. 42, p. 363, 1987.

[HAL 66] HALL K.R., EAGLETON L.C., ACRIVOS A. *et al.*, "Pore and solid-diffusion kinetics in fixed bed adsorption under constant pattern conditions", *Industrial & Engineering Chemistry Fundamentals*, vol. 5, p. 212, 1966.

[HAS 76] HASHIMOTO K., MIURA K., "A simplified method to design fixed-bed adsorbers for the Freundlich isotherm", *Journal of Chemical Engineering of Japan*, vol. 9, p. 388, 1976.

[HEL 52] HELBIG W.A., "Adsorption. Liquid phase processes are important", *Chemical Engineering*, vol. 59, pp. 153–157, 1952.

[HEL 67] HELFFERICH F.G., "Multicomponent ion exchange in fixed beds", *Industrial & Engineering Chemistry Fundamentals*, vol. 6, p. 362, 1967.

[HIE 52] HIESTER N.K., VERMEULEN T., "Saturation performance of ion-exchange and adsorption columns", *Chemical Engineering Progress*, vol. 48, p. 503, 1952.

[HIG 54] HIGGINS I.R., ROBERTS J.T., "A counter current solid–liquid contactor for continuous ion exchange", *Chemical Engineering Progress Symposium Series*, vol. 50, p. 87, 1954.

[HIG 64] HIGGINS I.R., "Use ion exchange when processing brine", *Chemical Engineering Progress*, vol. 60, p. 60, 1964.

[HOR 76] HORVATH C., MELANDER W., MOLNAR I., "Solvophobic interactions in liquid chromatography with nonpolar stationary phases", *Journal of Chromatography A*, vol. 125, p. 129, 1976.

[HUT 73] HUTCHINS R.A., "New method simplifies design of activated carbon systems", *Chemical Engineering*, vol. 20, p. 133, 1973.

[IGL 76] IGLESIAS H.A., CHIRIFE J., "B.E.T. monolayer values in dehydrated foods and food components", *Lebensmittel Wissenschaft und Technology*, vol. 9, p. 107, 1976.

[JUR 52] JURY S.H., LICHT W., "Adsorption wave in drying beds", *Chemical Engineering Progress,* pp. 102–109, 1952.

[KAC 02] KACZMARSKI K., CAVAZZINI A., SZABELSKI P. *et al.*, "Application of the general rate model and the generalized Maxwell–Stefan equation to the study of the mass transfer kinetics of a pair of enantiomers", *Journal of Chromatography A*, vol. 962, p. 57, 2002.

[KAS 81] KAST W., "Adsorption aus der Gasphase. Grundlagen und Verfahren", *Chemie Ingenieur Technik*, vol. 53, p. 160, 1981.

[KEL 56] KELLY N.W., SWENSON L.K., "Comparative performance of cooling tower packing arrangements", *Chemical Engineering Progress*, pp. 263–268, 1956.

[KEM 48] KEMBALL C., RIDEAL E.K., GUGGENHEIM E.A., "Thermodynamics of monolayers", *Transactions of the Faraday Society*, vol. 44, p. 948, 1948.

[KER 68] KERSCHNER E., BÖHNER G., SCHNEIDER A., "Beitrag zur wärmeübertragung bei der Furniertrocknung mit Düsenbelüftung", *Holz als Roh-und Werkstoff*, vol. 26, p. 19, 1968.

[KLE 67] KLEIN G., TONDEUR D., VERMEULEN T., "Multicomponent ion exchange in fixed beds", *Industrial & Engineering Chemistry Fundamentals*, vol. 6, p. 339, 1967.

[KNO 86] KNOX J.H., PYPER H.M., "Framework for maximizing throughput in preparative liquid chromatography", *Journal of Chromatography A*, vol. 363, p. 1, 1986.

[KRA 52] KRAMERS H., CROOCKEWIT P., "The passage of granular solids through inclined rotary kilns", *Chemical Engineering Science*, vol. 1, no. 6, pp. 259–265, 1952.

[KUB 65a] KUBIN M., "Beitrag zur Theorie der Chromatographic", *Collection of Czechoslovak Chemical Communications*, vol. 30, p. 1104, 1965.

[KUB 65b] KUBIN M., "Beitrag zur Theorie der Chromatographie Einfluss der Diffusion usserhalb und der Adsorption innerhalb des Sorbens–Korn", *Collection of Czechoslovak Chemical Communications*, vol. 30, p. 2900, 1965.

[LAP 52] LAPIDUS L., "Adsorption. Theory and practice are converging", *Chemical Engineering*, pp. 164–166, 1952.

[LAP 54] LAPIDUS L., ROSEN J.B., "Experimental investigations of ion exchange mechanisms in fixed beds by means of an asymptotic solution", *Chemical Engineering Progress Symposium Series*, vol. 50, p. 97, 1954.

[LEE 66] LEE H., CUMMINGS W.P., "A new design method for silica gel air driers under nonisothermal conditions", *Chemical Engineering Progress Symposium Series*, vol. 63, no. 74, p. 42, 1966.

[LEN 83] LENART A., FLINK J.M., "An improved proximity equilibration cell method for measuring water activity of foods", *Lebensmittel-Wissenschaft & Technologie*, vol. 16, no. 2, p. 84, 1983.

[LEV 41] LEVERETT M.C., "Capillary behaviour in porous solids", *Transactions of the AIME*, vol. 142, p. 152, 1941.

[LEW 22] LEWIS W.K., "The evaporation of a liquid into a gas", *Mechanical Engineering*, pp. 445, 1922.

[LIA 67] LIAPIS R.I., LITCHFIELD R.J., "Numerical solution of moving boundary transport problems in finite media by orthogonal collocation", *Computers & Chemical Engineering*, vol. 3, pp. 615–621, 1967.

[LON 85] LONCIN M., *Génie industriel alimentaire,* Masson, 1985.

[MAN 52] MANTELL C.L., "Adsorption. Bigger role lies ahead", *Chemical Engineering*, pp. 166–168, 1952.

[MEL 67] MELROSE J.C., "Exact geometrical parameters for pendular ring fluid", *The Journal of Physical Chemistry*, vol. 71, no. 11, pp. 3676–3678, 1967.

[MEL 03] MELCION J.P., HARI J.L., *Technologie des pulvérulents dans les I.A.A.*, Lavoisier, 2003.

[MER 25] MERKEL F., Verdunstungskühlung. (=Forschungsarbeiten auf dem Gebiete der Ingenieurwesens; Heft 275) und Zahlentafeln 1 bis 29, Verein deutscher Ingenieure, 1925.

[MER 52] MERIMS R., "Adsorption. Fixed bed design more empirical than moving bed", *Chemical Engineering*, pp. 161–163, 1952.

[MEY 67] MEYER O.A., WEBER T.W., "Non isothermal adsorption in fixed beds", *AIChE Journal*, vol. 13, p. 457, 1967.

[MIC 52] MICHAELS A.S., "Simplified method of interpreting kinetic data in fixed bed ion exchange", *Industrial & Engineering Chemistry*, vol. 44, p. 1922, 1952.

[MIC 78] MICHELSEN M.L., VILLADSEN J., "A convenient computational procedure for collocation constants", *Chemical Engineering Journal*, vol. 4, p. 64, 1978.

[MIN 96] MINAKUCHI H., NAKANISHI K., SOGA N. *et al.*, "Octadecylsilylated porous silica rods as separation media for reversed-phase liquid chromatography", *Analytical Chemistry*, vol. 68, p. 3498, 1996.

[MOL 23] MOLLIER, R., "Ein neues Diagram für Dampfluftgemische", *Zeitschrift der Verein deutschen Ingenieure*, vol. 67, p. 869, 1923.

[MOR 65] MORROW N.R., HARRIS C.C., "Capillary equilibrium in porous materials", *Society of Petroleum Engineers Journal*, vol. 5, p. 15, 1965.

[MOR 95] MORIN V., TRICHAIYAPORN S., STEINMETZ D. *et al.*, "Séchage du sucre en lit fluidisé", *Entropie*, vol. 191, p. 47, 1995.

[MOU 95] MOURAD M., HÉMATI M., LAGUERIE C., "Séchage du maïs en lit fluidisé à flottation Partie 1", *Chemical Engineering Journal*, vol 60, p. 39, 1995.

[MUJ 80] MUJUMDAR A.S., "Heat and mass transfer in granular porous media", *Advances in Drying*, vol. 1, pp. 23–61, 1980.

[NAD 95] NADEAU J.-P., PUIGGALI J.-R. *Séchage*, Lavoisier, 1995.

[NEI 82] NEISS J., *Numerische Simulation des Wärme-und Feuchte-transports und der Eisbildung in Böden*, Fortschritt-Berichte der VDI-Zeitschriften, 1982.

[NEL 56] NELSON R.L., VERMEULEN T., "Interpretation and correlation of ion exchange column performance under moulinear equilibria", *AIChE Journal*, vol. 2, p. 404, 1956.

[NOR 96] NORTON T.T., FERNANDEZ E.J., "Viscous fingering in size exclusion chromatography: insights from numerical simulation", *Industrial & Engineering Chemistry Research*, vol. 35, p. 2460, 1996.

[PAL 87] PALACHA Z., FLINK J.M., "Revised PEC method for measuring water activity", *International Journal of Food Science & Technology*, vol. 22, p. 485, 1987.

[PER 73] PERRY J.H. (ed.), *Chemical Engineers' Handbook*, 5th Edition, McGraw Hill, 1973.

[PIE 60] PIETSCH W., RUMPF H., "Haftkraft Kapillardruck, Flussigkeitsvolumen und Grenzwinkel einer Flussigkeitsbrücke zwischen zwei Kugeln", *Chemie Ingenieur Technik*, vol. 39, no. 15, pp. 885–893, 1960.

[POI 95] POINCARÉ H., *Capillarité*, Éditions Georges Carré, 1895.

[RAB 68] RABB A., "Are dry cooling towers economical?", *Hydrocarbon Processing*, vol. 47, no. 2, pp. 122–124, 1968.

[RAI 64] RAIMONDI P., TORCASO M.A., "Distribution of the oil phase obtained upon imbibition of water", *Society of Petroleum Engineers Journal*, vol. 4, p. 49, 1964.

[RAN 52] RANZ W.E., MARSHALL W.R., "Evaporation from drops", *Chemical Engineering Progress*, vol. 48, p. 141, 1952.

[RAZ 78] RAZAVI M.S., MC COY B.J., CARBONELL R.G., "Moment theory of breakthrough curves for fixed-bed adsorbers and reactors", *The Chemical Engineering Journal*, vol. 16, p. 211, 1978.

[REA 78] REAY D., *Proc. 1er Symposium international sur le séchage*, pp. 136–144, 1978.

[RHE 71] RHEE H., BODIN B.F., AMUNDSON N.R., "A study of the shock layer in equilibrium exchange systems", *Chemical Engineering Science*, vol. 26, p. 1571, 1971.

[RHE 72] RHEE H.K., AMUNDSON N.R., "A study of shock layer in non-equilibrium exchange systems", *Chemical Engineering Science*, vol. 27, p. 199, 1972.

[ROS 52] ROSEN J.B., "Kinetics of a fixed bed system for solid diffusion into spherical particles", *Journal of Chemical Physics*, vol. 20, p. 387, 1952.

[ROS 54] ROSEN J.B., "General numerical solution for solid diffusion in fixed beds", *Industrial & Engineering Chemistry*, vol. 46, p. 1590, 1954.

[ROU 61] ROUNSLEY R.R., "Multimolecular adsorption equation", *AIChE Journal*, vol. 7, no. 2, p. 308, 1961.

[RUT 80] RUTHVEN D.M., LEE L.-K., YUCEL H., "Kinetics of non-isothermal sorption in molecular sieve crystals", *AIChE Journal*, vol. 26, p. 16, 1980.

[RUT 81] RUTHVEN D.M., LEE L.-K., "Kinetics of nonisothermal sorption: systems with bed diffusion control", *AIChE Journal*, vol. 27, p. 654, 1981.

[RUT 85] RUTHVEN D.M., "Generalized statistical model for the prediction of binary adsorption equilibria in zeolites", *Industrial & Engineering Chemistry Fundamentals*, vol. 24, p. 27, 1985.

[SCH 68] SCHNEIDE P., SMITH J.M., "Adsorption rate constants from chromatography", *AIChE Journal*, vol. 14, p. 762, 1968.

[SIL 50] SILLEN L.G., "On filtration through a sorbent layer", *Arkiv för kemi*, vol. 2, p. 477, 1950.

[SNY 87] SNYDER L.R., COX G.B., ANTLE P.E., "A simplified description of H PLC. Separation under overload conditions. A synthesis and extension of two recent approaches", *Chromatographia*, vol. 24, pp. 82–96, 1987.

[SOL 67] SOLT G.S., "Continuous countercurrent ion exchange: the C.I. process", *British Chemical Engineering*, vol. 12, p. 516, 1967.

[SPI 92] SPIEGEL M.R., *Formules et tables de mathématiques*, McGraw-Hill, 1992.

[SUZ 71] SUZUKI M., SMITH J.M., "Kinetic studies by chromatography", *Chemical Engineering Science*, vol. 26, p. 221, 1971.

[THO 44] THOMAS H.C., "Heterogeneous ion exchange in a flowing system", *Journal of the American Chemical Society*, vol. 66, p. 1664, 1944.

[THO 48] THOMAS H.C., "Chromatography, a problem in kinetics", *Annals of the New York Academy of Sciences*, vol. 49, p. 161, 1948.

[THO 68] THOMPSON A.R., "Cooling towers", *Chemical Engineering*, pp. 100–102, 1968.

[TIM 69] TIMMINS R.S., MIR L., RYAN M., "Large scale chromatography: new separation tool", *Chemical Engineering*, vol. 76, p. 170, 1969.

[TIS 43] TISELIUS A., "Studien über Adsorptions analyse I", *Kolloid Zeitschrift*, vol. 105, p. 101, 1943.

[TON 67] TONDEUR D., KLEIN G., "Multicomponent ion exchange in fixed beds", *Industrial & Engineering Chemistry Fundamentals*, vol. 6, p. 351, 1967.

[TON 68] TONDEUR D., "Théorie des colonnes d'échange d'ions", *Chimie et Industrie–Genie Chimique*, vol. 100, p. 1058, 1968.

[TON 70] TONDEUR D., "Theory of ion-exchange columns", *Chemical Engineering Journal*, vol. 1, p. 337, 1970.

[TRE 52] TREYBAL R.E., "Adsorption. Techniques are ideal for difficult separations", *Chemical Engineering*, pp. 149–152, 1952.

[UCH 76] UCHIYAMA T., "Cooling tower estimates made easy", *Hydrocarbon Processing*, pp. 93–96, 1976.

[VAL 76a] VALENTIN P., "Determination of gas–liquid and gas–solid equilibrium isotherms by chromatography: I. Theory of the step-and-pulse method", *Journal of Chromatographic Science*, vol. 14, p. 56, 1976.

[VAL 76b] VALENTIN P., "Determination of gas–liquid and gas–solid equilibrium isotherms by chromatography. II. Apparatus, specifications and results", *Journal of Chromatographic Science*, vol. 14, p. 132, 1976.

[VAL 88] VALENZUELA D.P., MYERS A.L., TALU O. *et al.*, "Adsorption of gas mixtures: effect of energetic heterogeneity", *AIChE Journal*, vol. 34, p. 397, 1988.

[VAS 83] VASSEUR J., Doctoral thesis in engineering, ENSIA, Massy, 1983.

[VER 52] VERMEULEN T., HIESTER N.K., "Ion exchange chromatography of trace components", *Industrial & Engineering Chemistry*, vol. 44, p. 636, 1952.

[VER 54] VERMEULEN T., HIESTER N.K., "Ion exchange and adsorption columns kinetics with uniform partial presaturation", *The Journal of Chemical Physics*, vol. 22, p. 96, 1954.

[VIL 67] VILLADSEN J.V., STEWART W.E., "Solution of boundary-value problems by orthogonal collocation", *Chemical Engineering Science*, vol. 22, p. 1483, 1967.

[VIL 69] VILLERMAUX J., VAN SWAAIJ W.P.M., "Modèle représentatif de la distribution des temps de séjour dans un réacteur semi-infini à dispersion axiale avec zones stagnantes. Application à l'écoulement ruisselant dans des colonnes d'anneaux Raschig", *Chemical Engineering Science*, vol. 24, p. 1097, 1969.

[WAK 58] WAKAO N. OSHIMA T., YAGI S., "Mass transfer from particles to fluid in packed beds", *Kagaku Kogaku*, vol. 22, p. 780, 1958.

[WAL 45] WALTER J.E., "Multiple adsorption from solutions", *The Journal of Chemical Physics*, vol. 13, p. 229, 1945.

[WHI 77a] WHITAKER S., "Toward a diffusion theory of drying", *Industrial and Engineering Chemistry Fundamentals*, vol. 16, no. 4, pp. 408–414, 1977.

[WHI 77b] WHITAKER S., "Simultaneous heat, mass and momentum transfer in porous media: a theory of drying", *Advances in Heat Transfer*, vol. 13, pp. 119–203, 1977.

[WHI 83] WHITAKER S., CHOU W.T-H., "Drying granular porous media theory and experiment", *Drying Technology*, vol. 1, no. 1, pp. 3–33, 1983–1984.

[WHI 91] WHITLEY R.D., VAN COTT K.E., BERNINGER J.A. *et al.*, "Effects of protein aggregation in isocratic nonlinear chromatography", *AIChE Journal*, vol. 37, p. 555, 1991.

[WIL 40] WILSON J.N., "A theory of chromatography", *Journal of the American Chemical Society*, vol. 62, p. 1583, 1940.

[WIL 45] WILKE C.R., HOUGEN O.A., "Mass transfer in the flow of gases through granular solids extended to low modified Reynolds numbers", *Transactions of the American Institute of Chemical Engineers*, vol. 41, p. 445, 1945.

[WIL 66] WILSON E.J., GEANKOPLIS C.J., "Liquid mass transfer at very low Reynolds numbers in packed beds", *Industrial & Engineering Chemistry Fundamentals*, vol. 5, p. 9, 1966.

[YOU 67] YOUNG J.H., NELSON G.L., "Theory of hysteresis between sorption and desorption isotherms in biological materials", *Transactions of the American Society of Agricultural Engineering*, vol. 10, pp. 260–263, 1967.

[YU 89] YU Q., WANG N.-H.L., "Computer simulations of the dynamics of multicomponent ion exchange and adsorption in fixed beds. Gradient-directed moving finite element method", *Computers & Chemical Engineering*, vol. 13, p. 915, 1989.

Index

W, V